Disposal of Hazardous Waste in Underground Mines

The Sustainable World

Aims and Objectives

Sustainability is a key concept of 21st century planning in that it broadly determines the ability of the current generation to use resources and live a lifestyle without compromising the ability of future generations to do the same. Sustainability affects our environment, economics, security, resources, health, economics, transport and information decisions strategy. It also encompasses decision making, from the highest administrative office, to the basic community level. It is planned that this Book Series will cover many of these aspects across a range of topical fields for the greater appreciation and understanding of all those involved in researching or implementing sustainability projects in their field of work.

Topics

Data Analysis
Data Mining Methodologies
Risk Management
Brownfield Development
Landscaping and Visual Impact Studies
Public Health Issues
Environmental and Urban Monitoring
Waste Management
Energy Use and Conservation
Institutional, Legal and Economic Issues
Education
Visual Impact
Simulation Systems
Forecasting
Infrastructure and Maintenance
Mobility and Accessibility
Strategy and Development Studies
Environment Pollution and Control
Land Use
Transport, Traffic and Integration
City, Urban and Industrial Planning
The Community and Urban Living
Public Safety and Security
Global Trends

Main Editor

Associate Editors

Disposal of Hazardous Waste in Underground Mines

Editors

V. Popov
Wessex Institute of Technology, UK

R. Pusch
Geodevelopment AB, Germany

WIT*PRESS* Southampton, Boston

Disposal of Hazardous Waste in Underground Mines

Series: The Sustainable World, Volume 11

Editors

V. Popov
Wessex Institute of Technology, UK

R. Pusch
Geodevelopment AB, Germany

Published by

WIT Press
Ashurst Lodge, Ashurst, Southampton, SO40 7AA, UK
Tel: 44 (0) 238 029 3223; Fax: 44 (0) 238 029 2853
E-Mail: witpress@witpress.com
http://www.witpress.com

For USA, Canada and Mexico

WIT Press
25 Bridge Street, Billerica, MA 01821, USA
Tel: 978 667 5841; Fax: 978 667 7582
E-Mail: infousa@witpress.com
http://www.witpress.com

British Library Cataloguing-in-Publication Data

A Catalogue record for this book is available
from the British Library

ISBN: 978-1-85312-750-2
ISSN: 1476-9581

Library of Congress Catalog Card Number: 2006921658

Printed in Great Britain by Lightning Source UK Ltd., Milton Keynes.

Contents

Chapter 2

Chapter 3

Chapter 5

Preface: Towards a safer future

This book summarizes the ideas and concepts of a group of European business and academic centres from the UK, Sweden, Germany and Greece on disposal of hazardous waste (HW) in underground mines. This collaboration started as a result of a joint multidisciplinary European effort sponsored by the European Commission as part of the Energy, Environment and Sustainable Development Programme under the project title Low Risk Disposal Technology (LowRiskDT). The developed theoretical concepts and practical application technologies were further advanced by the group since the conclusion of the LowRiskDT Project, and the main findings of this joint work are summarized in this book.

The research work was focused on: the possibility of using abandoned underground mines for disposal of hazardous chemical waste with negligible pollution of the environment; the properties and behaviour of waste-isolating clay materials and practical ways of preparing and applying them; development of software tools to assess the stability, performance and risks associated with different repository concepts, considering the long-term safety of the biosphere; the isolating capacity of reference repositories; and the different approaches for handling hazardous chemical waste. The project has demonstrated that HW can be safely disposed in underground mines provided that adequate assessment, planning and design procedures are employed.

Abandoned mines are sites where advanced exploration or mining activities ceased without rehabilitation having been implemented at all or completed. There are many abandoned mines in the EU where environmental and economic benefits would exist if those mines are used for disposal of HW. Disposal of HW in abandoned underground mines would in many cases reduce environmental risks from pollution after sealing the mine workings. Industry and waste management companies will benefit from this alternative approach by gaining access to a safe and relatively cheap strategy for dealing with HW.

Since the assessment of suitable mines must be done on a case-by-case basis, this book cannot specify a certain type of mine as suitable for safe disposal of HW. However, research has led to the definition of the interdisciplinary scientific approach needed to identify the suitable mine types, and also the technology needed in order to achieve safe disposal of HW in underground mines. The results of a comparison between underground storage of HW in abandoned mines, the use of landfills for disposal of HW and other suitable alternative ways are reported.

The issue of abandoned mines, with the associated physical, environmental and public safety concerns, constantly emerges around the world as a reminder of the legacy that past mining operations have created. Some abandoned mines give rise mainly to physical concerns. These concerns include public health and safety, visual impacts, stability issues and dust problems. Accidents related to vertical openings or deteriorating structures are the most common cause of death and injury in abandoned mines. Lethal concentrations of explosive and toxic gases like methane, carbon monoxide and hydrogen sulphide can accumulate in underground passages. Rock falls and cave-ins from adits or pit walls can be safety hazards. Unsafe structures include support timbers, ladders, cabins and other related features.

Abandoned mines and associated features can also have a detrimental effect on soils, water, plants and animals. The extent of the effects is not fully known because inventories are incomplete and some efforts are still being evaluated.

Water is one of the resources most frequently polluted by abandoned mines. Water is also the main conduit by which impacts from abandoned mines extend beyond the immediate site. Elevated concentrations of metals and increased levels of suspended sediment, acidity, hydrocarbons, and brine leaching can threaten surface and underground water quality and aquatic habitats.

The socioeconomic consideration of abandoned mines arises mostly from both the physical and the environmental effects discussed in the preceding paragraphs. These include the safety hazards caused by abandoned mines that can result in the loss of lives.

Contamination of both ground and surface waters by abandoned mines impacts on the aquatic systems, which affects communities that depend on fishing for their livelihoods. There may also be imposition of restrictions on legitimate users of the waters who may find it unsuitable for irrigation, livestock watering, industrial or domestic use.

Funds are required for the rehabilitation of abandoned mine sites. The question when dealing with abandoned mines is: who provides these funds, what mechanisms exist in various jurisdictions to raise these funds, and who is ultimately responsible for the rehabilitation work and the long-term care of the sites. In some cases governments are forced to take on the task of rehabilitation when there are no identifiable owners or if the owners have no resources to pay for rehabilitation.

Some mines are situated so that present or past mining operations have caused contamination of the groundwater. For such mines backfilling and sealing for isolation of stored HW can in fact lead to improved environmental conditions.

The situation in Europe is defined by the EC Landfill Directive (Council Directive 1999/31/EC of 26 April 1999 on the Landfill of Waste, Commission of the European Communities, *Official Journal of the European Communities*, 16/7/1999, L 182/1-19), which was adopted in July 1999. It sets out new operational, regulatory and technical requirements for the landfilling of waste. In the past a large amount of HW was co-disposed with non-hazardous waste (NHW) but this practice had to end. The overall effects of the additional measures within the Directive are to: (i) increase disposal costs, (ii) prohibit the long established practice of co-disposing HW with NHW, and (iii) require the development of additional treatment facilities. Some analysis (Implementation of Council Directive 1999/31/EC on the Landfill of Waste – Second Consultation Paper, Department for Environment, Food and Rural Affairs, London, August, 2001) indicates that hazardous liquid and solid wastes are likely to incur high per ton additional costs.

Part of the EC acceptance criteria for HW (Section 2.4 of Commission Decision establishing criteria and procedures for the acceptance of waste at landfills pursuant to Article 16 and Annex II of Council Directive 1999/31/EC on the Landfill of Waste, Commission of the European Communities, Brussels, 01.05.2002) which specifies the leaching limit values for different components of the HW, applies only for landfills for HW and not for underground storage facilities. Therefore, disposal of HW in underground storage facilities would reduce the requirements for treatment of the hazardous solid wastes.

There are indications that adaptation of some of the existing abandoned underground mines would be cheaper than building new landfill repositories for disposal of HW, especially taking into account the long-term monitoring of landfills for HW and pollution prevention measures.

In many countries disposal of inorganic waste that can contaminate groundwater is presently made in the form of landfills equipped with clay-based top and bottom liners. The longevity of the liners is limited because of erosion of the top liner and chemically induced degradation of the bottom liner, particularly if the pH of the percolate is low. The continuous percolation of the waste mass by rain and meltwater means that the ion-sorbing capacity of the bottom liner is ultimately exhausted so that no further sorption can take place after a certain period of time. The percolate then passes through the bottom liner to the groundwater with the same concentration of contaminating ions as the porewater has in the lower part of the waste mass. By collecting and cleaning the percolate, pollution of the groundwater can be prevented for some period of time; but it is questionable whether the institution responsible will continue to take care of the cleaning operation for decades and centuries and to safely dispose the concentrate that results from the cleaning process.

For many of the waste materials that are presently being accumulated on the ground surface, disposal in abandoned mines would offer an alternative option implying very slow or no percolation at all for a very long time perspective. Hence, the only transport mechanism is diffusion; but if solid waste is mixed with even a small amount of expandable clay, diffusive migration of hazardous ions released will be very much delayed. The most important gain achieved by mine disposal is that the long-term impact on the environment will be much smaller than for disposal on the ground surface, and there are also other important benefits such as saving the ground surface for more valuable purposes than waste disposal.

Disposal of waste, such as incinerated ash, low-level radioactive waste and contaminated soil, using suitable simple versions of the clay-isolation principle may not represent higher cost than common landfilling, and the capitalized cost over a long period of time should be lower since no maintenance or monitoring will be required.

Ash commonly has, or can be given, a gradation that is suitable for mixing with fine-grained expandable clay-like powder of Friedland Ton. Past experience tells that the dry density can be raised sufficiently to give the mixture a low hydraulic conductivity and a low ion diffusion capacity as well as a certain expandability. Placement of such mixtures in drifts and rooms in abandoned mines using modern backfilling and effective compaction techniques thus means that the clayey ash can not only be effectively isolated from the biosphere but also support can be provided for the rock to avoid convergence of the backfilled rooms and hence settlement of the ground surface, which is a common problem in mined-out areas.

It is also proposed that low-level radioactive waste (LLW) with short-lived radionuclides can be disposed in mines, since it requires guaranteed isolation of the waste for a few hundred years. Considering the fact that suitably designed and constructed clay barriers with a thickness of less than 1 m will not even be water-saturated in this period, as shown by our research, the disposal concept described in this book should be acceptable. The high risk of groundwater contamination by geological disposal of waste that can give off radionuclides would still require careful selection of suitable mines and particularly effective isolation potential of the clay barriers.

An attractive version of disposal of LLW in mines, for both environmental and cost reasons, would be to apply it in freshly and carefully excavated drifts and rooms in actively used mines while continuing the mining operation in other parts.

Where environmental hazards have taken place, of which the Chernobyl nuclear accident is a typical example, large amounts of soil have to be dug out and transported to suitable sites for cleaning or isolation. Other examples are areas where impregnation of arsenic in wood has taken place or where a chemical industry has been located and caused contamination of the soil with heavy metals, pesticides and similar compounds. The quantities of soil that have to be removed can be hundreds of thousands of cubic meters, for which bigger abandoned mines will be suitable. The same principles have to be followed as for disposal of ash and similar preparation of mixtures of waste in particulate form with fine-grained expandable clay powder, preferably using the sandwiching method.

Considering the above, a number of suitable mines should be prepared and characterized with respect to the type of contaminated soil that they can store in case of unexpected events.

The Editors
2006

CHAPTER 1

Hazardous waste generation and management in Europe

D. Kaliampakos[1], A. Mavropoulos[2] & M. Menegaki[1]
[1]*School of Mining & Metallurgical Engineering, National Technical University of Athens, Greece.*
[2]*EPEM, Greece.*

Abstract

This chapter provides an overview of the hazardous waste generation and management strategies in Europe. The results show that four waste streams require special attention: (i) waste from the waste management industry; (ii) waste from organic chemical processes, more specifically pesticides, due to both their environmental impacts and their importance for the chemical industry; (iii) old batteries; and (iv) the waste from electrical and electronic waste stream. Some issues such as general correlation between GDP and waste from energy production, waste import to EU member states, and the concept of waste hierarchy for waste management are discussed. The relative environmental impact of waste is estimated by using information on the quantity and the degree of hazard associated with it. Waste with a high specific environmental impact per ton is normally found in minor volumes and is therefore more difficult to be separated and collected. Until now, waste management has mainly concentrated on waste streams in the middle of the area marked.

1.1 Introduction

The last 10 years have been characterized by major changes in hazardous waste generation and management. Changes in generation rates are not uniform in European countries and they strongly depend on the phase of economic development and the specific industrial sectors that characterize this phase. Some well-developed countries (e.g. Austria) have shown a great increase in hazardous waste generation (62%) while others (e.g. Denmark) have shown a substantial decrease (57%). On the other hand, in Ukraine hazardous waste generation has decreased by 38% between 1996 and 2000, while in the Russian Federation quantities have increased by 32% between 1996 and 1999. Such dramatic differences are connected with the legislation and the definition of hazardous waste in each country. The lack of a uniform definition for hazardous waste and consequently the absence of a common classification system result in questionable conclusions regarding the total quantities and the general trends in hazardous waste generation and management. However, in terms of quality there are certain issues that should be taken into consideration.

Several studies indicate that a limited number of specific industrial sectors contribute substantially to hazardous waste generation in each country. An EEA study [1] has shown that a large proportion of hazardous waste in most Western European countries consists of a relatively small number of waste types (typically 75% of hazardous waste consists of 20 principal types, a very small number compared to the 236 different hazardous list codes). The major types differ from one country to another, but in most EU countries hazardous waste generation is dominated by a relatively small number of sources. This means that prevention and recycling efforts can be relatively easily focused at specific industries in each country, allowing the possibility of achieving remarkable results within a short period.

1.2 Hazardous waste generation in Europe

The EEA member countries generate about 36 million tons of hazardous waste per year [2]. It is extremely difficult to interpret the statistical data on hazardous waste. The analysis of the data shows large changes in reported amounts over time, as illustrated in Table 1.1. Countries and regions with figures for both 1990 and 1995 show an apparent increase (65% on average) in hazardous waste quantities, mainly due to changed definitions and new legislation. Germany, with figures for 1990 and 1993, and UK, with figures for 1990 and 1994, show a decline by an average of 21% before the introduction of the Hazardous Waste List (HWL). This decline can be possibly explained by the introduction of cleaner technologies.

The dissociation of waste generation from economic growth is a big challenge, since the detailed analysis of the relationship between those two factors reveals several different trends. For instance, country comparisons show no general correlation between GDP and waste from energy production, which probably reflects national differences in energy supply systems. Coal-fired power plants generate large amounts of fly ash, while hardly any waste is produced from hydroelectric power stations and nuclear power plants generate a small, but extremely hazardous, amount of waste. For hazardous waste a correlation between GDP and waste quantities can be demonstrated according to the data from 1995. Nevertheless, this is not valid for the data from 1990. In this period large changes took place both in awareness of hazardous waste and in definitions and classification procedures. Thus, the apparent correlation in 1995 may be spurious.

Table 1.1: Reported quantities (in tons) of hazardous waste in selected countries and regions [1–5].

	Year					
Country	1990	1993	1994	1995	1996*	1999*
Austria	317,000			577,000		
Denmark	106,000			252,000		
Netherlands			895,000			
Germany (North Rhine – Westphalia)*		1,597,671				
Germany	13,079,000	9,093,000				
UK	2,310,000		2,080,000			
Spain (Catalonia)	674,400			831,439		
Spain (Basque Country and Catalonia)					1,362,317	
Greece						286,856

*Quantities according to the first two digits of HWL codes.

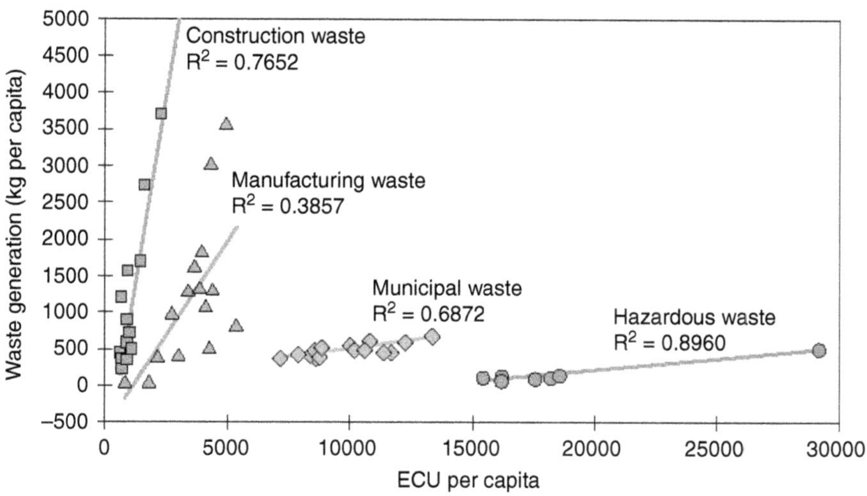

Figure 1.1: Correlation between waste generation and GDP per capita [2, 3, 6, 7].

For each member state, waste quantity per capita has been plotted against economic activity related to selected waste streams. Figure 1.1 shows that the generation of municipal, construction and hazardous waste seems to be related with the economic activity behind waste generation whereas such a relation does

Table 1.2: Total generation of hazardous waste in EEA member countries stated in tons. Generation per capita in kg [2, 4, 6, 8].

Country	Year	Total generation	Generation per capita	Classification
Austria	1992	423,000	54	National
	1994	513,000	64	National
	1995	577,000	68	National
Denmark	1990	106,000	21	National
	1995	250,000	48	National
Germany	1993	9,100,000	113	National
Ireland	1992	99,000	28	Basel
	1995	248,000	70	HWL

not seem to exist for manufacturing waste. A good correlation is assumed if R^2 values are above 0.7. In relation to municipal waste the economy is stated as final consumption from households in purchasing power standard (PPS). Hazardous waste is related to GDP stated in PPS. Construction and manufacturing waste are related to the part of the GDP originating from construction and manufacturing activities.

About 1,665,500 tons of hazardous waste were imported to EU member states and Norway in 1995. The majority of the above waste (approximately 85%) came from other EU member states, 8% came from other OECD countries, in particular Switzerland, US, Norway, Hungary and the Czech Republic, while the sources for an amount of about 6% are unknown. Many non-OECD countries do not have adequate facilities to treat their hazardous waste in a safe way. Until these countries are properly equipped, the EU could help by importing and treating their hazardous waste. However, only 16,000 tons (approximately 1%) of the imports to EU member states and Norway were hazardous waste from non-OECD countries, in particular from South Africa, Brazil, Macedonia and Slovenia. Some more details are provided in Table 1.2.

1.2.1 Hazardous waste generation per employee

The industrial structure varies within each country and region. The relative size of manufacturing industries is approximately the same in Denmark and Austria, while its importance in the Basque Country and Germany is greater, when measured by the number of full-time employees. The hazardous waste generated per employee in the manufacturing and other sectors of the countries, for which data were available, is given in Table 1.3.

The number of employees in different industrial sectors can explain the difference in quantities of hazardous waste generated only to a limited extent. Nevertheless,

Table 1.3: Hazardous waste generated (in kg) per employee in Austria, Denmark, Germany, Ireland and Spain (the Basque Country and Catalonia) according to NACE code [1, 9].

Country	Manufacture	Trade, service, transport and infrastructure
Austria (1996)	210	223
Denmark (1996)	144	135
Germany (1993)	372	129
Ireland (1996) – excluding recovery on site	436	628
Spain/Basque (1994)	888	49
Spain/Catalonia (1996)	156	1348

the generation of waste seems to be more closely related to the number of employees than to the total population. A few dramatic differences in waste per employee may, for certain industries, be explained by the presence of waste types considered hazardous in only one country.

1.2.2 Main waste streams in Europe

Almost 80% of the hazardous wastes become from eight activities as outlined in Fig. 1.2. More specifically, waste from the waste management industry and inorganic waste from thermal processes make up more than 36%. Waste from organic chemical processes, inorganic waste from metal treatment and coating of metals, and hazardous waste from construction and demolition have an equal contribution of almost 9% each.

Figure 1.3 presents the main waste streams, classified according to their percentage in the top five HWL 6-digit codes. For this purpose, more detailed profiles (based on 6-digit codes of HWL) of several countries have been used. These countries include Austria, Denmark, Ireland, Germany and Spain.

According to the data presented in Fig. 1.3 it is clear that:

- There is a lack of classified data (according to the HWL) in several countries. Even when there are available data, the use of different classification systems complicates the quantification of the trends in all the countries and regions. This issue has been officially considered as one of the major policy barriers for the development of a European level hazardous waste management.
- It is certain that hazardous waste amounts have increased, but it is difficult to quantify the rate of increase, due to lack of relevant data or due to no compatibility. For Austria, Spain/Catalonia and Denmark, the data show increasing quantities of hazardous waste in the period 1993–96 with higher rates of increase for Austria and Spain/Catalonia and smaller rates for Denmark.

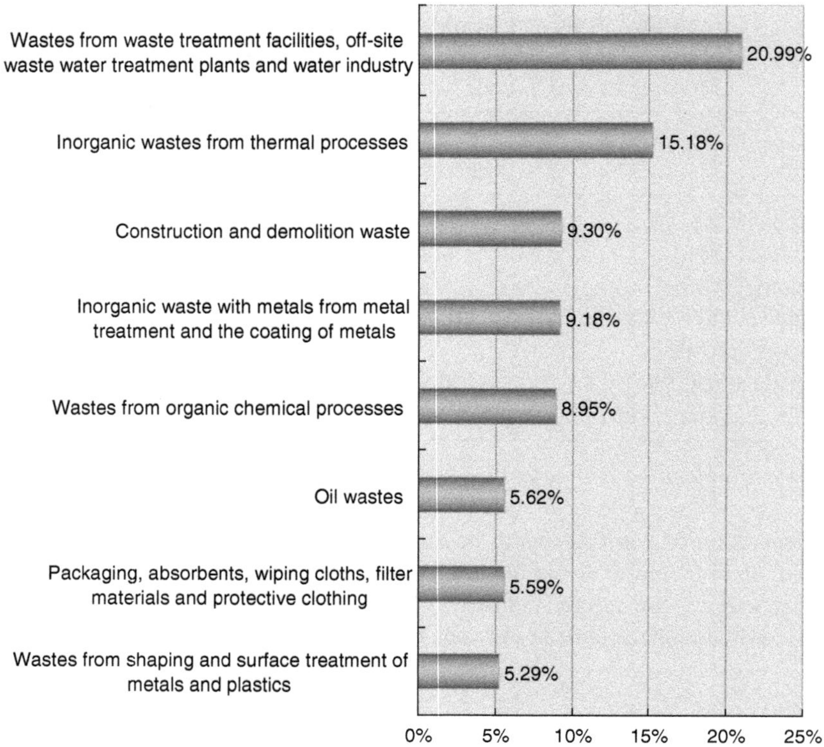

Figure 1.2: Main hazardous waste streams in EEA countries.

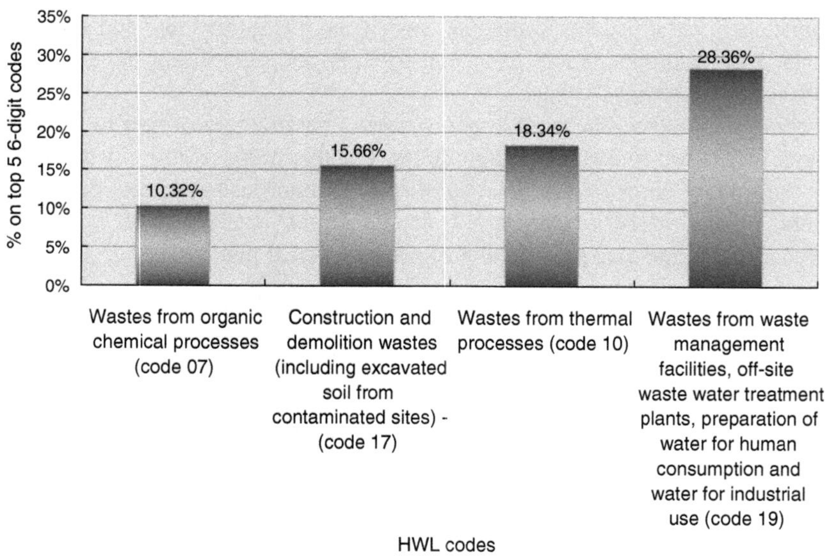

Figure 1.3: The main waste streams included in the top five HWL 6-digit codes.

- However, increasing amounts of hazardous waste can be the result of positive developments such as better collection and registration of waste and not necessarily as a result of a real increase in waste generation.

1.3 Current hazardous waste management in Europe

The EU has established the well-known concept of waste hierarchy for waste management. According to this concept, waste prevention and minimization is the only viable long-term solution for waste management, while recycling is the second preferred option. However, these undoubtedly right options must translate into a need to design materials, goods and services in such a way that their manufacture, use, reuse, recycling and end-of-life disposal results in the least possible waste. Thus, this concept requires great changes in economy,market and social behaviours and such changes need time to be prepared and applied.

In the mean time, hazardous waste management in Europe is characterized by great differences from country to country. In several Western European countries the main option is recovery of hazardous waste, while in the majority of the EU and European countries landfill or incineration without energy recovery are widely used. In many countries, hazardous waste has to be stabilized before disposal, using an appropriate physicochemical treatment. However, treatment methods are often poorly defined, sometimes they are even undeclared, leading to difficulties in comparing practices and environmental impacts. Tables 1.4 and 1.5, taken from Eurostat [10], provide the latest year available information regarding hazardous waste management activities in Europe.

Incineration of hazardous waste is a commonly used practice for disposal in many countries. According to the latest year available data from Eurostat [8], at least 4.72 millions tons of hazardous waste is incinerated without energy recovery, an amount comparable with 5.9 millions tons of hazardous waste that are treated by physicochemical methods. Social acceptance of incineration is a frequent problem, especially in the cases where local conditions eventually prohibit the sustainability of operations of the incineration plants (e.g. long transport routes). Although the specific technique can reduce the after-treatment residue of waste, not all hazardous waste are suitable for safe incineration. Moreover, flue-gas cleaning has become a very difficult and very expensive issue, especially for hazardous waste incinerators, after the release of the new EU directive for incinerators.

One further important issue is that part of the residue (fly ash and bag-filters) is hazardous waste and needs, in any case, another disposal option. It should be noted that slag and fly ash from waste incineration are two of the major hazardous waste streams in a number of Western European countries.

Landfilling of hazardous waste is officially considered the lowest-ranking waste management option. However, it still is the dominant method of disposal in Europe. According to the latest year available data, more than 13.2 million tons of hazardous waste is landfilled, an amount remarkably larger than the sum

Table 1.4: Recovery operations in hazardous waste management in Europe.

			Recovery operations			
Country	Year	Total	Incineration (energy recovery)	Recycling and composting	Other recovery operations	Preparatory activities
Belgium	1999	634				
Denmark	1998	224	156	68		
Germany	–					
Greece	1997	100				
Spain	1996		168	1,197	208	
France	1998			222		
Ireland	1998	153	5			
Italy	1997		92	629	159	
Luxemburg	1998					
Netherlands	1998	227				
Austria	–					
Portugal	–					
Finland	1997	61	42	19		92
Sweden	–					
UK	1993	196				
Iceland	1999	6				
Norway	1998		119			
Switzerland	1998	73				
Bulgaria	1997	317				
Czech Republic	1999		37	28	316	
Estonia	1999	78				
Hungary	1996	365				
Poland	1999	400				8
Romania	1999	414	3	411		
Slovak Republic	1998		68	158		

of all the other hazardous waste management techniques (11.8 millions tons). Environmental problems, as well as the reluctance among the public to accept landfills as a safe technology, make the establishment of new landfills extremely difficult. In most countries, hazardous waste landfills' capacities are very limited or unavailable. Thus, pending the availability of treatment and disposal options, hazardous waste is accumulating. Some countries (e.g. Estonia, Latvia) have demonstrated some success in this regard by establishing safe storage for large quantities of obsolete pesticides. However, this cannot be considered as a final solution. The need for an environmentally sound alternative to landfill is more than urgent. Figures 1.4 and 1.5 provide a condensed picture of current practices in hazardous waste management in Europe, as taken by Eurostat.

Table 1.5: Disposal–treatment operations in hazardous waste management in Europe.

			Disposal operations				
Country	Year	Total	Physico-chemical treatment	Biological treatment	Incineration (without energy recovery)	Landfill	Other
Belgium	1999				129	631	
Denmark	1998	57				57	
Germany	–						
Greece	–						
Spain	1996		686		750	189	
France	1998		302		1,361	803	
Ireland	1996	88	7		46	33	
Italy	1997			1,132	282	791	620
Luxemburg	1998				–	–	
Netherlands	1998		557		244	370	
Austria	1996				106		
Portugal	–						
Finland	1997	365	59	3	59	234	10
Sweden							
UK	1993		620		185	931	
Norway	1998		335		–		
Switzerland	1998		277		371	219	
Bulgaria	1998					237	
Czech Republic	1999		1,071	128	5	217	576
Estonia	1999		7		0	5,748	6
Hungary	1996		1,015		1,110	1,035	
Poland	1999	734				113	
Romania	1999	1,759	416		5	1,318	436
Slovak Republic	1998		592	103	68	392	25

1.4 Trends and expectations

1.4.1 Dangerous substances from waste streams and EU priorities

In order to predict future trends, it is more than necessary to take into account not only the distribution of hazardous waste but also the relevant contribution of dangerous substances from current waste management practices as well as EU priorities with regard to waste streams.

There are three types of emissions that are of relevance at the global level, namely:

- organic micropollutants, particularly dioxins and furans (incineration is still a major generating source);

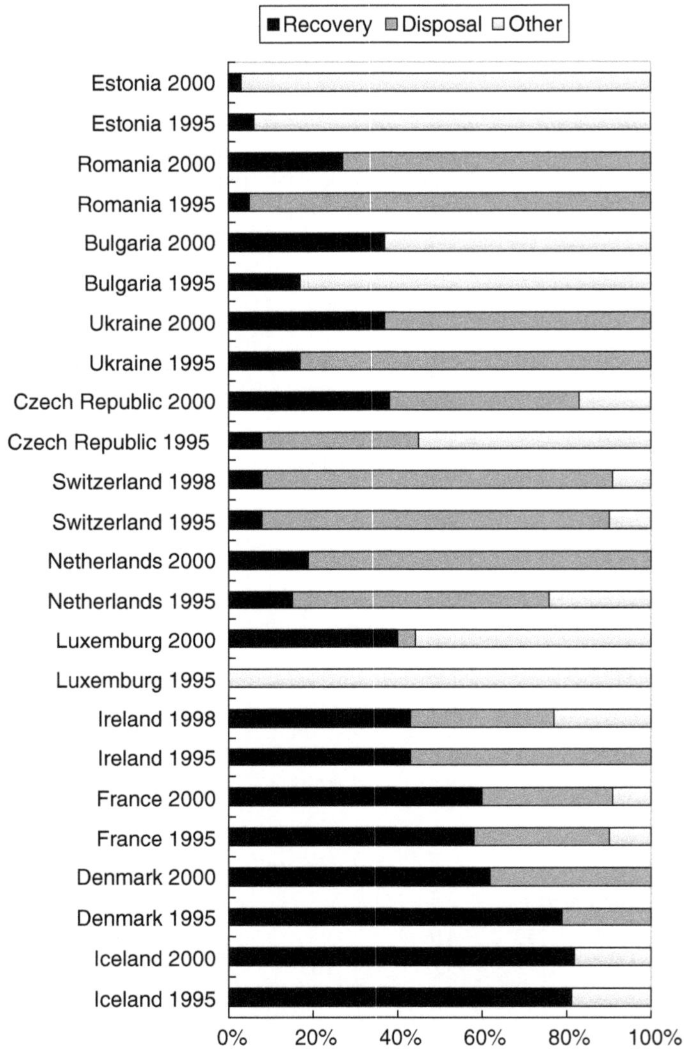

Figure 1.4: Current hazardous waste management in selected European countries.

- greenhouse gases, particularly methane (landfilling is one of the most important sources as stated in the EMEP/CORINAIR Atmospheric Emission Inventory Guidebook);
- volatile heavy metals (incineration is still a major generating source for specific metals).

Emissions of the above substances contribute to a slow but continuous degradation of environmental conditions. Other emissions from incineration, such as polychlorinated biphenyls (PCBs), have a more regional character and are important at the regional/local level.

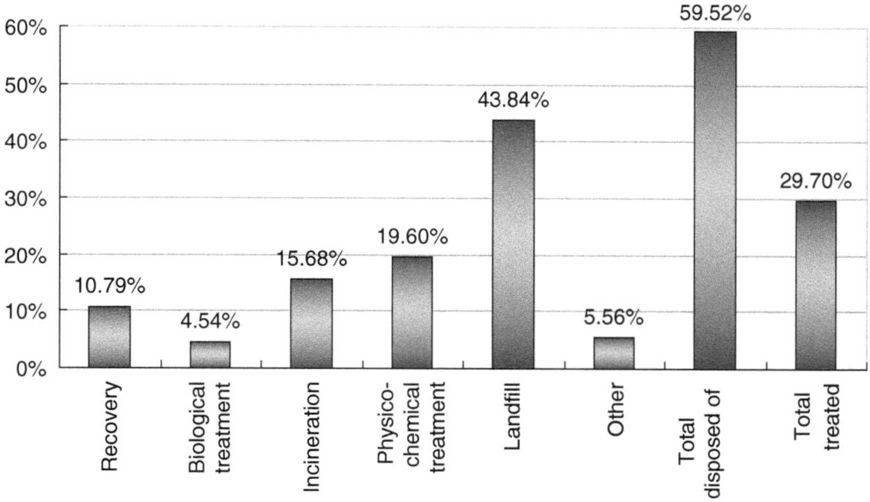

Figure 1.5: Current hazardous waste management practices in Europe.

Landfill emissions other than methane are mainly of local or regional importance. Most of these emissions are emitted in a diffuse manner to the surrounding environment and, in particular, to groundwater. In regions where communities rely on groundwater for public water supply, such emissions, if uncontrolled, can have implications for public health.

Organic trace substances produced as a result of biodegradation processes can also be a source of nuisance to local communities as well as a potential risk to human health.

For both landfill and incineration, discharge of wastewaters results in relatively high emissions of chloride salts.

Table 1.6 [11] provides some interesting remarks regarding the emissions from incineration and landfilling of selected waste streams.

In relation to this, it is interesting to note the results of a survey carried out by OECD [12] where 10 European countries were asked about their present and future waste minimization problems and priorities. According to the results obtained (Table 1.7) one of the most important waste streams is waste from electrical and electronic equipment (WEEE).

The production of electrical and electronic equipment is one of the fastest growing domains of manufacturing industry in the Western world. New applications of electrical and electronic equipment are increasing significantly. There is hardly any part of life where electrical and electronic equipment are not used. This development leads to an important increase in WEEE. The WEEE stream is a complex mixture of materials and components. In combination with the constant development of new materials and chemicals having environmental effects, this leads to increasing problems at the waste stage.

Table 1.6: Ranking of dangerous substances from landfill and incineration.

Dangerous substances	Source	Category	Remark
Organic compounds PCDD/PCDF	Incineration	Human toxicity Ecological toxicity	Very important, incineration is a major contributor
CH_4	Landfill	Greenhouse Gas	Very important
Volatile heavy metals Hg, Cd, Pb	Incineration	Human toxicity Ecological toxicity	Very important due to transboundary movement
Total N, NH_4	Landfill	Eutrophication	Important because of the local contamination of surface and groundwater
HCl	Incineration	Acidification	Important
Heavy metals As, Cd	Incineration	Human toxicity	Important, carcinogenicity
Salt, e.g. chloride	Landfill Incineration	Ecological toxicity	Important, high loads to surface and groundwater
Organic emissions	Landfill	Human toxicity Nuisance	Important for employees and neighbourhood
Heavy metals Cd, Ni, Cu, Zn, Pb, Hg	Landfill	Ecological toxicity Human toxicity	Less important due to the small contribution to total emissions, assumed to be stable in the landfill body

The WEEE stream differs from the municipal waste stream for a number of reasons:

- The rapid growth of WEEE. In 1998, 6 million tons of WEEE were generated (4% of the municipal waste stream). The volume of WEEE is expected to increase by at least 3–5% per annum. This means that in 5 years 16–28% more WEEE will be generated and in 12 years the amount would have doubled. The growth of WEEE is about three times higher than the growth of the average municipal waste (WEEE Draft Directive).
- Because of their hazardous content, electrical and electronic equipment cause major environmental problems during the waste management phase if not properly pretreated. As more than 90% of WEEE is landfilled, incinerated or recovered without any pretreatment, a large proportion of various pollutants found in the municipal waste stream comes from WEEE.
- The environmental burden due to the production of electrical and electronic products ('ecological baggage') by far exceeds the environmental burden due

Table 1.7: Present and future key waste streams in selected European countries.

Country	Present key waste streams	Future key waste streams
Austria	Waste oil, lubricants, photochemical	WEEE, waste medicines, end-of-life vehicles
Denmark	No information	No information
Finland	Sewage sludge	WEEE, end-of-life vehicles
France	Waste oil, end-of-life vehicles	Waste oil, end-of-life vehicles, PCB, WEEE, medical waste
Germany	Paint sludge, WEEE, sewage sludge	WEEE, end-of-life vehicles, sewage sludge
Italy	No information	No information
Netherlands	Waste oil, dredging spoil, CD waste, phosphorogypsum	Dredging spoil, phosphorogypsum
Norway	Hazardous waste in general	WEEE, scrapped oil installations
Switzerland	Packing, beverage containers, metal plating sludge	Packing, beverage containers, metal plating sludge
UK	WEEE, end-of-life vehicles, waste oils	Clinical waste, PCB

to the production of materials constituting the other substreams of the municipal waste stream. As a consequence, enhanced recycling of WEEE should be a major factor in preserving resources, in particular, energy.

In order to address adequately the environmental problems associated with the current methods for the treatment and disposal of WEEE, it is considered appropriate to introduce measures at the Community level that aim, firstly, at the prevention of WEEE, secondly at the reuse, recycling and other forms of recovery of such wastes and, thirdly, at minimizing the risks and impacts to the environment from the treatment and disposal of WEEE.

Although the main guideline for WEEE management is separate collection and reuse, recycling or recovery, there is a large part of WEEE, concerning old and not rechargeable batteries, which may need extended disposal, at least for some years, until some new and more effective technologies are well established.

Since 1990, mercury consumption in primary batteries has declined significantly in the EU due to the introduction of the Directive 91/157/EEC on batteries and accumulators containing certain dangerous substances. The Directive came into force in 1994. The Directive covers, amongst the other types of batteries, the commonly used alkaline-manganese energy cell, the zinc–carbon battery, the zinc–air button cell as well as the silver oxide button cell and the mercuric oxide battery (two small battery types which also contain mercury). Tables 1.8 and 1.9 present mercury emissions from batteries and other sources.

Table 1.8: Emission factor for mercury: tons per million population.

Source	Western Europe	Eastern Europe
Batteries	0.0002	0.00004
Measurement and control equipment	0.0044	0.0013
Electrical equipment	0.0019	0.0003
Lighting	0.0005	0.0003

Table 1.9: Mercury emissions in Europe (in tons) (1995).

Source	Western Europe	Eastern Europe
Batteries	0.09	0.015
Measuring and control equipment	1.81	0.500
Electrical equipment	0.77	0.125
Lighting	0.21	0.121
Total	2.88	0.761

1.4.2 Future trends up to 2010

The use of certain chemicals is expected to decline over the next decade in the EU. However, a growth of 30–50% in chemicals output is expected for most of the EU countries by 2010 as a result of increasing economic activity, including road transport and agricultural production (Table 1.10). This anticipated growth could accentuate concerns with respect to human and ecosystem health. Considerable uncertainties exist over both the projections for emissions (and consequently concentrations and depositions levels) and the relationships between exposure and effects; emission uncertainties for dioxin, for example, range from 5 to 20. Nevertheless, it is important to consider the future trends for major groups of persistent chemicals due to the potential risk of significant impacts. Atmospheric emissions, concentrations and depositions have been modelled [13] on a European scale for selected heavy metals, persistent organic pollutants (POPs) and for fine particulate matters (PM10).

Emission estimates for 1990 have been prepared within the framework of the joint OSPARCOM–HELCOM–UNECE emission inventory [14] and are used to construct projections for the year 2010.

1.4.3 Emission trends of heavy metals

Lead emissions from phasing out leaded petrol (85/210/EEC) have been reduced more than 50%, on average, in the EU and the Accession Countries between

Table 1.10: Drivers for chemical use and resulting exposure.

Main drivers	Chemical	Source
Food production	Pesticides, Cd, Hg	Crop treatment, phosphate fertilizers, seed treatment
Transport	Pb Pt, Pd, PAHs organics	Petrol additives (in some countries); catalytic converters; incomplete combustion oil refining
Fuel conversion	PAHs, Cu, Cd, Hg, As	Incomplete combustion, fly ash
Mining, metals industry	Cu, Cd, Hg, As	Ore processing, zinc refining
Consumer goods and products (GDP growth)	Dioxins, furans	Waste incineration

1990 and 1996 and further reductions are expected by 2010. The concurrent introduction of catalytic converters, however, will most likely result in increased platinum emissions, either through direct release or in the course of reprocessing.

Projections indicate that positive trends from abatement measures, such as improved efficiency and geographical coverage in recycling, are likely to be counteracted by a general increase in economic activity [13]. Thus, the overall cadmium and mercury emissions are expected to increase in EEA countries by 26% and 30%, respectively, between 1990 and 2010. Copper emissions (mainly from mining and smelting activities) have increased by 8% and are unevenly distributed between countries. Policies in the pipeline lead to an appreciable decrease in emissions of lead, copper and mercury in the Accession Countries, although cadmium emissions are expected to increase by about 4% due to an increase in road transport and growth of the chemical industry.

Heavy metal and arsenic emissions are expected to be diminished as a result of the reduction of the sulphur content of fuels [following EU legislation COM(97)88] and the switch from solid to liquid fuels [15], which are frequently associated with pyrite, the main sulphur source in coal and lignite. Although the improvements in wastewater treatment techniques and the degree of water treatment connections, as well as the tighter controls on industrial discharges, have led to reduced heavy metal river loads, they have intensified the problem of contaminated sludge disposal.

1.4.4 Emission trends of pesticides and POPs

Increases in general economic activity, including agricultural production, are projected to counteract positive trends from abatement measures [13]. Policy measures in the framework of the Integrated Pollution Prevention and Control

(IPPC) directive and its predecessors are expected to reduce emissions of dioxins/furans (from large-scale combustion plants) and of PCBs. Measures aimed at reducing energy consumption and/or improving conversion efficiency are also expected to have positive effects. In Western Europe the anticipated growth in road transport is expected to increase polycyclic aromatic hydrocarbons (PAHs, for instance, benzo[*a*]pyrene) and xylene emissions. In the Accession Countries this will be offset by the introduction of cleaner vehicles, although an expected increase in the annual volumes of incinerated waste will lead to increased emissions of hexachlorobenzene (HCB).

1.5 The effect of Directive 99/31

The European Directive 99/31 raises environmental standards for all types of landfills (municipal, inert and hazardous waste landfills) and pushes hard to wards a big increase of the already high landfill tipping fees. The specific directive imposes stringent operational and technical requirements on landfilling and calls for the reduction in the quantity of the various waste streams entering the landfills as well as for the treatment of all waste prior to landfill. The main scope of this directive is to eliminate landfills. The application of the Landfill Directive is usually accompanied by landfill taxes that increase much more the cost of landfilling.

The combination of the new Landfill Directive and the environmental and social problems that characterize landfills has led to a remarkable reduction of landfills from a number of almost 10,000 (1991, 12 countries of Central Europe) to almost 5,000 (1999). Landfill space has become much more limited, landfilling is politically driven to be much more expensive and landfill is considered as the less-preferred option in the EU. Figure 1.6 shows the consequent reduction of landfills in Europe.

One very interesting issue is that the Landfill Directive makes a distinction between underground storage and landfill. According the directive, underground storage means a permanent waste storage facility in a deep geological cavity, such as a salt or potassium mine. On the other hand landfill means a waste disposal site for the deposit of the waste on to or into land. Article 3 excludes underground storage facilities from:

- The provisions of Article 13, paragraph d. The specific paragraph concerns the need for environmental monitoring and aftercare of landfill sites for about 20 years after closure and stoppage of their operation.
- The provisions of Annex 1, point 2, which specifies stringent obligation for leachate management in landfills (collection, treatment, protection of surface water).
- The provisions of Annex 1, points 3–5, which specify certain measures for protection of soil and water, such as liners, gas handling and control, etc.
- The provisions of Annex 2 that mention specific waste acceptance criteria for landfills.

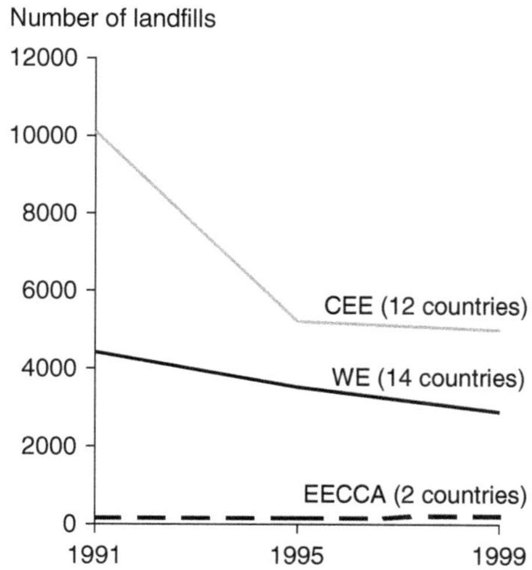

Figure 1.6: Reduction of landfills in Europe [16, 17]. (WE: Austria, Belgium, Finland, France, Ireland, Italy, Luxembourg, Portugal, Spain, Sweden, the Netherlands, Iceland, Norway, Switzerland; CEE: Croatia, Cyprus, Czech Republic, Estonia, Hungary, Latvia, Lithuania, Malta, Poland, Romania, Slovak Republic, Turkey; EECCA: Belarus, Tajikistan)

Finally, Article 16 and Annex 2 underline the need for the development of specific waste acceptance criteria for underground storage. The exclusion of underground storage from many of the strict obligations that should be applied for landfills results in a big advantage for underground disposal. While capital and operational cost of landfills have become remarkably higher, in accordance with new, highly technical and environmental standards, the construction and operational standards for underground disposal are much more easy to achieve and the related costs are substantially cheaper compared to landfill costs.

1.6 Waste streams and pollutants of interest

1.6.1 Waste streams of interest

A general outline of the hazardous waste profiles in several EU countries has already been shown. Based on these data, as well as on the EU priorities, it is clear that the key waste streams as far as hazardous waste is concerned are the following:

- Waste coming from the waste management industry is the main hazardous waste substream. The biggest waste stream is the aqueous liquid waste from gas treatment and other aqueous liquid waste. This waste includes a number of heavy metals and the physical form of this waste makes it appropriate for storage within artificial barriers.

- Although waste from construction and demolition activities and inorganic waste from thermal processes are of major interest, waste from organic chemical processes should be also considered as a main hazardous waste sub-stream due to their environmental importance. This type of waste has long-term interactions with the environment, its pollutants are bioaccumulative and part of this waste represents the greatest problem of the disposal of outdated pesticides.
- WEEE is one of the key future waste streams. This stream also includes a number of heavy metals and there are many difficulties in recovering materials due to lack of appropriate technological advances. Since it is considered a certainity that within the next 5–10 years appropriate technologies will be established, underground mines can provide a perfect medium–term storage option, with very low cost and negligible environmental hazards. The same conclusion can be drawn regarding the old batteries stream, as a part of the rapid increasing (and EU special focus) waste stream of WEEE.

In all the above cases, underground mine disposal takes advantage of the very slow groundwater flow at depth and the very long transport paths for heavy metals that might be released.

1.6.2 Pollutants of interest

After the decision-making of waste streams of interest, the next step to be taken is the selection of specific pollutants. In order to do this, it is necessary to provide a general outline regarding the dispersion of chemical substances (pollutants) and its effects.

1.6.3 Selection of the pollutants of interest

Europe is one of the largest chemical-producing regions in the world, supplying 38% of the global turnover. Since 1993, the chemical intensity of EU GDP has been rising for all chemicals and hazardous chemicals production. There are 20,000–70,000 thousand substances or groups of substances in the European market, many being derived from chlorine-based organic chemistry. Little is known about the toxicities, eco-toxicities or risks from most of these substances.

Figures on the quantities of substances produced or marketed are in general of little use for predicting the dispersion and the potential exposures, which are still difficult to estimate due to increasing non-point sources of emissions and recycling processes, despite the improvements in multimedia modelling.

The European coverage of monitoring data for halogenated organics in general and for POPs in particular is rather patchy. Information on degradations, transformations, by-products and exposures to mixtures is also poor. Most monitoring programme focus on mobile media (air, water), but often neglect soil, sediments and consumer products.

Combustion of fossil and other organic fuels is thought to account for over 90% of the environmental load of the 280 types of carcinogenic PAHs.

Emissions of dioxins, such as polychlorinated dibenzo-*p*-dioxins (PCDD) and polychlorinated dibenzofurans (PCDF), which are mainly from air emissions and waste, are significant, but falling in most countries.

As mentioned before, if current trends and policies continue, there could be a growth of 30–50% in chemicals output for most of the EU countries by 2010.

Cadmium and mercury emissions are expected to increase by 26% and 30%, respectively, between 1990 and 2010; some countries plan to phase out these substances. Emissions from pesticides and POPs – such as dioxins/furans and PCBs – will continue to decrease, but PAH, HCB and xylene emissions are likely to increase.

However, the impact of some emerging trends in the management of chemicals such as eco-efficiency improvements, a shift from products to services, the internalization of external environmental costs into prices via taxes, increased information to the public, increased evidence on low-dose effects, greater use of the precautionary principles, and implementation of the OSPAR/Sintra agreement, the IPPC directive and other international policies could lead to marked reductions in the chemical intensity of European GDP, particularly for those substances of concern.

In order to focus on pollutants coming from organic chemical processes, there is a need to have a clearer idea about the chemical substances dispersed in Europe.

1.6.4 Chemical substances

Humans and ecosystems are constantly exposed to a mixture of natural and manufactured chemicals, some of which are not necessarily harmful. The 'chemicals intensity' [18] and 'dangerous chemicals intensity' of the EU economy (production plus imports per unit of GDP) [19] have been increasing since 1993 (Fig. 1.7). Supplying 38% of global turnover, with Western Europe being responsible for 33%, Europe is one of the largest chemical-producing regions in the world and the industry is expected to continue its vigorous growth. In line with the decline in their GDP between 1989 and 1995, the chemical production in the Accession Countries has been decreased; nevertheless it has seen a recovery in more recent years. The higher value-added brands of the sector are expected to flourish, benefiting from increases in both domestic demand and exports. However, the social and environmental costs of harmful environmental and health impacts are difficult to quantify [20] and rarely borne by those responsible for these impacts.

Human and ecosystem exposure depend upon the dispersion pattern of chemicals, which is determined by their physicochemical properties, their respective release modes, the environmental medium into which they are first released [21], their reactivity and degradability, and the kinetics of these physical and chemical processes. Certain chemicals, most notably chemical elements, never degrade, while organic substances may have half-lives and environmental residence times ranging from a few days up to geological timescales. Assessment of dispersion and exposure is exceedingly difficult; while processes can be studied in a

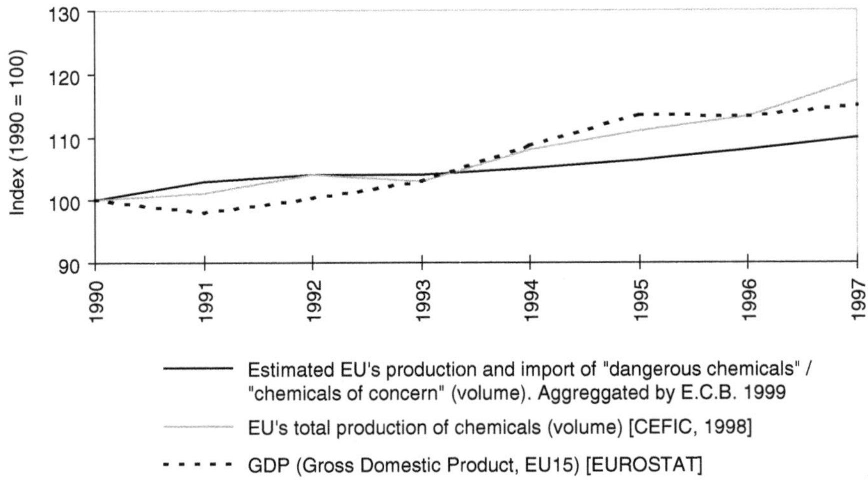

Figure 1.7: EU production and import of 'dangerous chemicals'/'chemicals of concern' and total chemical production.

laboratory, their impact in varied environmental conditions is very uncertain. Data on environmental impacts can only be gathered through extensive and continuing monitoring of both the environmental concentrations of selected substances and their impacts on environmental compartments. A recent workshop on environmental monitoring [22] has emphasized the importance of long time series in order to be able to detect changes with time, but has also highlighted the need to re-evaluate older data with improved scientific insight. Data on the biotoxicity and toxicity of chemicals is very limited (Fig. 1.8).

For 75% of large-volume chemicals (marketed in excess of 1000 tons a year) there is insufficient publicly available data even for minimal risk assessment under the OECD guidelines [18].

Figure 1.9 provides a very useful picture in order to clear up the situation regarding our knowledge for these substances. Currently up to 70,000 or more chemical substances are used for different purposes. A significant number of these substances do not occur naturally, but are manufactured, in some cases, in large quantities (HPVCs), with a resulting high statistical probability of human exposure. Many of the HPVCs are used in a vast range of manufactured goods and other products that are considered essential for modern life, including detergents and other 'down-the-drain' chemicals. Several hundred new substances are marketed each year and are recorded on the European Lists of Notified Chemical Substances (ELINCS), which has over 2000 substances listed.

The European Inventory of Existing Commercial Chemical Substances (EINECS) lists over 100,000 substances which are supposed to have been on the market in 1981. However, only 10,000 are produced in volumes greater than 10 tons/year. There is very little data on the dispersion or fate or effects of most substances.

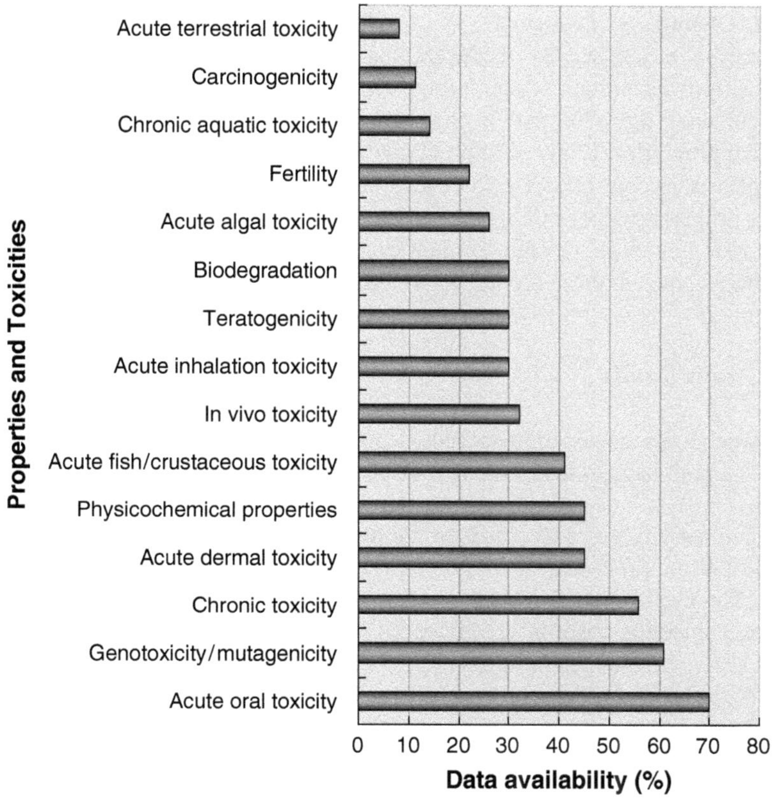

Figure 1.8: Availability of data on 2472 high-production volume chemicals (HPVCs) submitted to the European Chemicals Bureau [23].

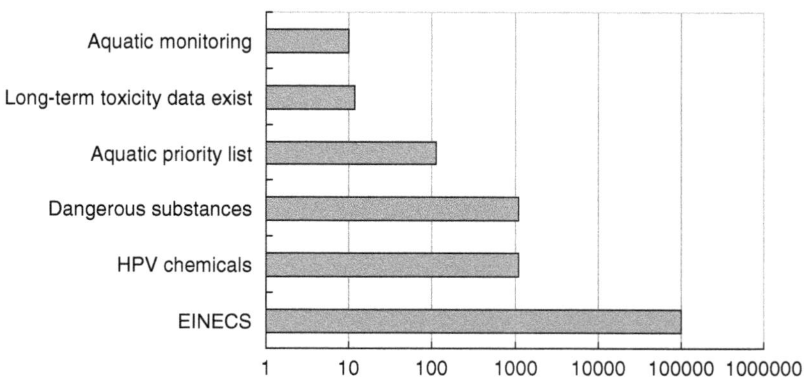

Figure 1.9: The chemical universe in contrast with some current monitoring and classification activities.

1.6.4.1 Chemicals of concern

The analysis is focused on chemicals that can be driven to water resources via leakage from hazardous waste storage facilities. The migration behaviour of a chemical along the water path is largely determined by its chemical and physico-chemical properties. Redistribution of contaminants between the aqueous and the solid phases is controlled by the prevailing water chemistry and the resulting surface properties. Zones of high acidity or alkalinity (measured by low/high pH) and/or redox potential or high sorption capacity (e.g. clays) can act as geochemical barriers. Indeed, these properties are utilized in engineered landfills to prevent leakage.

1.6.5 Heavy metals

Heavy metals are metals or metalloids, which are stable and have a density greater than 4.5 g/cm^3, namely lead, copper, nickel, cadmium, platinum, zinc, mercury and arsenic.

Human activity generally leads to the dispersion of metals and other elements that have been concentrated by geological processes and over geological timescales. The use of metals, as well as human exposure to metals, has significantly increased since the onset of the industrial revolution and continues to do so on a global scale. Arsenic, cadmium, copper, lead and nickel have been identified as being of the greatest concern.

The production and use of heavy metals is driven by a wide variety of industrial, agricultural and domestic uses such as metallurgy, catalysts, pigments for paints, batteries, electronics components, fertilizers, solid fossil fuels, plastics and fuel additives. The major diffuse anthropogenic mercury source, in Germany for instance [24], is the burning of fossil fuels. EU average contributions from agriculture to cadmium emissions are around 1%. Cadmium in phosphate fertilizers is of some concern [25] and is dealt with by national legislation in Finland, Sweden and Austria.

1.6.5.1 Heavy metals in water pathways

Direct human exposure to elevated heavy metal concentrations via the water pathway has been of limited importance in many Western European areas, but has regained importance as a result of declining control over the quality of groundwater resources and the distribution system. This may, for example, increase human lead exposure from drinking water, countering the lead solubility control measures for water piping. Exposure to surface water-derived heavy metals might occur indirectly via bioaccumulation in freshwater or estuarine or marine fish consumed by humans. The latter, for instance, accounts for half of the mercury intake in Germany [24]. The increasing abundance of biological wastewater treatment plants throughout Europe leads to a shift of environmental dispersion pathways of heavy metals from effluents to sludges. Sludges are either used as fertilizers (if contaminant concentrations are within legal limits) or are incinerated. River loads have decreased considerably as a result of wastewater treatment [14]. The LOIS studies [26] have confirmed that elevated concentrations of heavy metals in river waters are linked to the presence of high-suspended particle loads and natural or anthropogenic complexants. Heavy metals remobilized from stream sediments are of some concern where drinking water resources are augmented by bank filtration. The ultimate recipients of heavy metals in surface waters are the large marine basins.

1.6.5.2 Heavy metals in soil

High heavy metal concentrations in soils tend to be more localized, either from high natural background levels (mineral deposits), or from mining, ore processing and other industrial activities.

1.6.5.3 Heavy metals in food

The major pathway for human uptake, after inhalation, is ingestion of plant- and animal-derived foodstuffs. The chemical processes associated with bioaccumulation, both in humans and animals, lead to preferential accumulation in certain tissues. Wahlström *et al.* [27] concluded that the consumption of fish in Finland can be considered, in general, safe, but their liver and kidneys should be avoided. Human exposure to heavy metals may not only result from dietary uptake but also from smoking.

1.6.6 Persistent organic pollutants

The number of chemicals characterized as POPs is unknown, but certainly exceeds those that are listed as important [28, 29] or are included in current monitoring programmes. The term persistent organic pollutants includes 'chemical substances that persist in the environment, bioaccumulate through the food web and pose a risk of causing adverse effects to human health and the environment' [29].

1.6.6.1 Polycyclic aromatic hydrocarbons

PAHs comprise a suite of around 280 substances from which 16 have been selected by the EU and the US EPA as priority substances [30, 31]. PAHs are ubiquitous and many have environmental half-lives in excess of weeks or months. They are subject to various chemical and photochemical processes in the environment; some of which result in degradation to less toxic products, while others lead to more hazardous compounds, such as nitrosubstituted PAHs [32]. The major sources of PAHs are fossil and other organic fuels such as wood. Combustion is thought to account for over 90% of the environmental concentrations. Non-combustion processes such as the production and use of creosote and coal tar, though poorly quantified, are potentially very significant primary and secondary sources [30]. Combustion processes have the highest dispersion potential over wide areas, but may significantly decrease as emissions are reduced by IPPC measures, although total emissions are liable to increase with economic activity.

Human exposure occurs mainly through inhalation of smoke particles to which the PAHs readily attach. Indeed, certain voluntary practices such as smoking and the use of household chemicals such as air fresheners result in significant indoor PAH concentrations and human exposures [33].

1.6.6.2 Organochlorines dispersal in soil, groundwater and some global-scale problems

Chlorine-based organic chemistry has become one of the most important branches of the chemical industry [34] accounting for approximately 55% of the production. The main products are pesticides and biocides as well as components for a wide range of industrial and household goods. A class of chlorinated hydrocarbons released into the natural environment on purpose is those that are intended as plant protection products (insecticides, fungicides, herbicides) and biocides. Active ingredients may be not only chlorinated hydrocarbons but also other organic, metal–organic or metal compounds. Application generally results in a diffuse source of emissions (e.g. from agriculture, or organo-tin anti-fouling paints on ships), but for specific applications linear (e.g. weed control on railways) or point sources (wood protection, sheep dips, accidental spills) may be relevant.

Emission factors in industrial applications and household goods vary considerably, but are generally quite low during normal use; there are small losses from the technosphere by means of abrasion, wear or leakages, notably PCBs from electrical installations. PVC-based plastics have been of some concern, mainly due to emissions of additives, such as stabilizers or plasticizers (e.g. phthalates, chlorinated paraffins), in the waste stream and from consumer goods intended for children's use. The recycling of many PVC-based goods and the better process control in incineration has reduced the impact from dioxin formation in thermal waste treatment.

Most lipophilic organochlorines (those that are absorbed by fats) are found in the soil solids (the organic or clay fraction), from where they can migrate into deeper strata. A number of European countries have reported pesticides in groundwater; although there is little reliable information for POPs in general. The pollutants eventually reach the sea via surface waters and, inter alia, by colloid or particle-mediated transport.

Acute poisoning of humans by chlorinated hydrocarbons is rare in the European region and usually associated with accidental releases during manufacturing, storage or application. Bioaccumulation [34] and persistence in many environmental media can lead to long-term low-level exposure of non-target species. Health effects on humans and animals from continuous or intermittent long-term exposure to low levels are varied and frequently difficult to attribute. Certain pathological observations, including eggshell thinning in various bird species, skeleton malformation in seals and otters, hormonal (endocrine) or reproductive disturbances in various species, were found to coincide with pesticides in the animal tissue [35].

The continuing use of some active ingredients of concern, for instance DDT in developing countries, results in dispersive input to European regions, even though the respective ingredients have been phased out in Western Europe [29]. Lower acute human toxicity and easier handling for less well-educated farmers might be valid reasons for continuing their use in developing countries [36]. The ever-increasing global trade in plant material (foodstuffs, textiles) provides another, anthropogenic, pathway for transboundary dispersion and possible human exposure in Europe.

The overall pesticide use – measured by mass of active ingredient – appears to have been decreasing in most EU countries over the past two decades (Figures 1.10 and 1.11) [37].

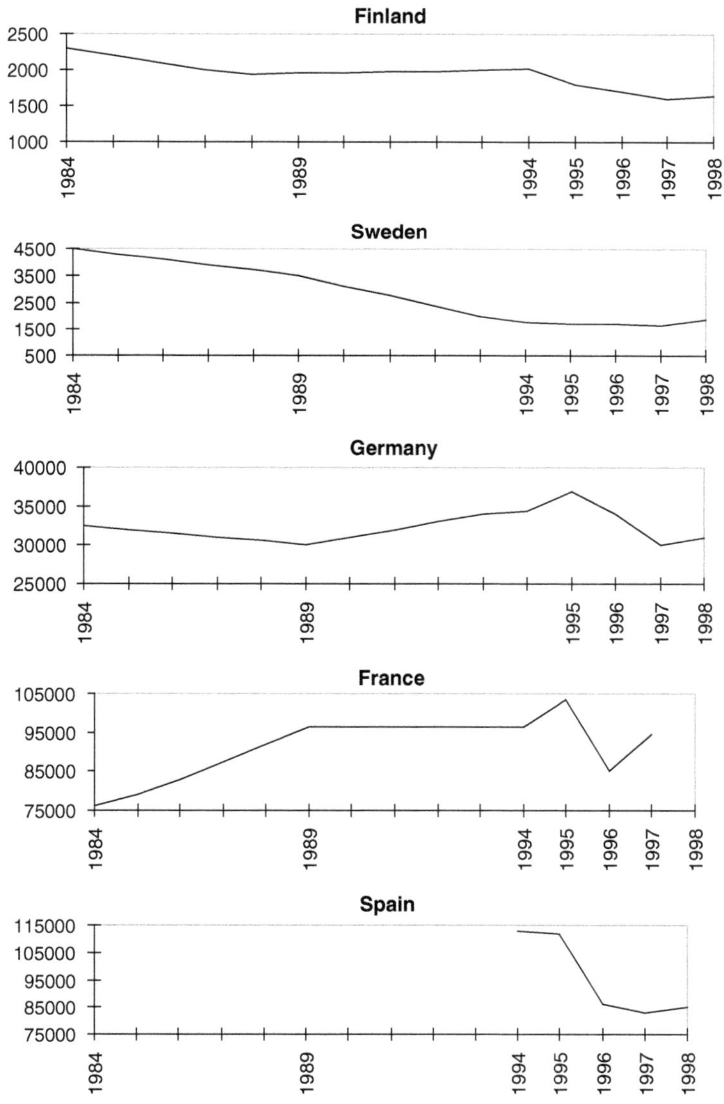

Figure 1.10: Pesticide consumption in selected countries of EU (in tons of active ingredients).

As shown in Fig. 1.11, pesticide consumption in the various EU countries does not follow a uniform trend, being a function of agricultural activity and legislation of certain substances. Absolute levels reflect the size of the countries well as the respective importance of the agricultural sector. Consumption in terms of mass, however, does not necessarily reflect environmental impact, as more active and more specific substances are being developed.

As far as the DDTs and lindanes are concerned, their production and use have already been reduced or prohibited a long time ago. Nevertheless, it will take considerable time for the reservoirs in various environmental compartments to become depleted and for stockpiles to be exhausted.

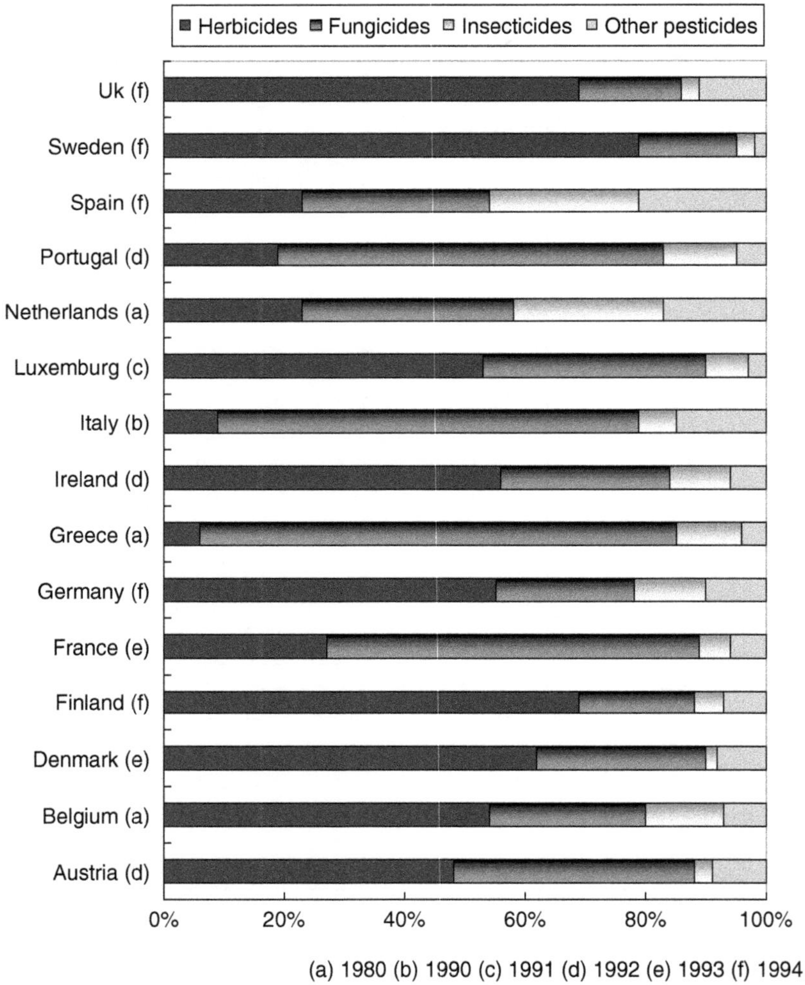

Figure 1.11: Percentage consumption of pesticides according to their types.

1.7 Conclusions

From the analysis presented above it is concluded that four waste streams require special attention:

- Waste coming from the waste management industry.
- Waste from organic chemical processes. More specifically pesticides should be considered as a waste stream of major importance, due to both their environmental impact and their importance in the chemical industry.
- Old batteries.
- The rapid increasing (and EU special focus) waste stream of WEEE.

The relative environmental impact of waste is related to both the quantity and the degree of hazard associated with it. There are, therefore, two aspects as far as waste generation is concerned: (i) quantitative, i.e. how much is generated, and (ii) qualitative, i.e. the degree of hazard. This is presented for a selection of materials in Fig. 1.12. Waste with a high specific environmental impact per ton is normally found in minor volumes and is therefore more difficult to be separated and collected. Until now, waste management has mainly concentrated on waste streams in the middle of the area marked.

Figure 1.13 ensures that heavy metals and pesticides are of major importance, due to their enormous environmental impact.

Finally, regarding the selection of pollutants of high importance, two pollutant categories have been selected.

First, heavy metals coming from the waste management industry waste, such as incineration slag or from old batteries. The contribution of batteries to heavy metal emissions has already been mentioned before. Some more details are given hereinafter for waste coming from the waste management industry.

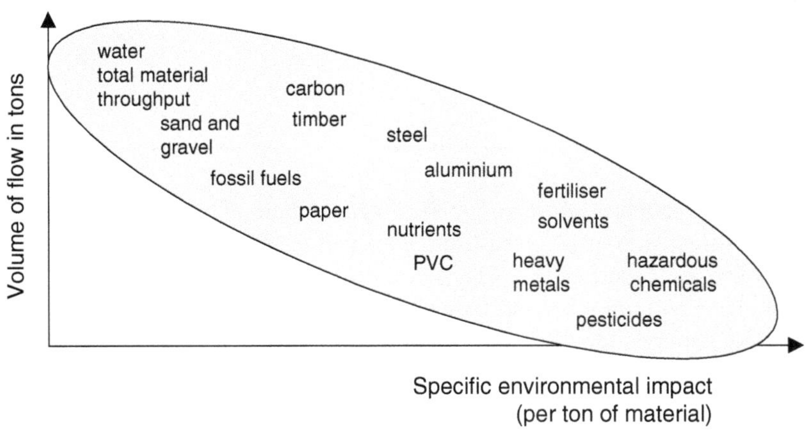

Figure 1.12: Environmental impacts per ton of waste materials [38].

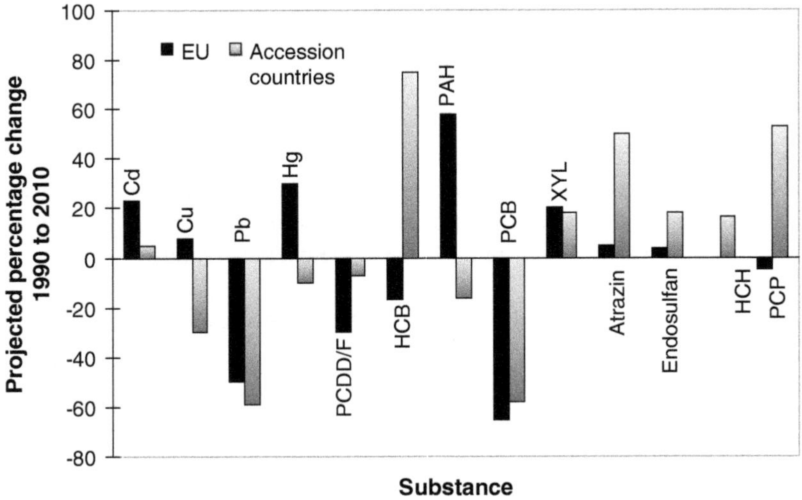

Figure 1.13: Projected percentage changes (1990–2010) in emissions of selected chemicals.

Table 1.11: Comparison between slag and soil parameters.

	Concentrations (mg/l)		
	Range in slag	Range in natural soils	Dutch value for good soil quality
As	0.12–189	1–50	29
Hg	0.02–7.75	0.01–0.3	0.3
Cd	0.3–70.5	0.01–0.70	0.8
Cr	23–3.170	1–1000	100
Cu	190–8.240	2–100	36
Ni	5–500	7–4.280	35
Pb	98–13.700	2–200	85
Zn	613–7.770	10–300	140
PAH	13–19.000		1

Based on available information, the total amount of slag from incinerator plants is estimated to be between 6 and 9 million tons per year in EEA countries. In a number of countries the slag is recycled and used for road construction, embankments and noise barriers as well as for concrete production. In Denmark and the Netherlands the percentage of slag that is recycled is between 85% and 90%, while only 50% is recycled in Germany and hardly any slag is recycled in Sweden [39, 40]. When analysing the chemical composition of incinerator slag a major concern is the heavy-metal content which in many cases is consider ably higher than the concentrations occurring naturally in soil (Table 1.11).

This means that in many cases the use of slag for construction purposes may in the long-term lead to contamination of surrounding areas with dust containing heavy metals, if the surface is not sealed. On the other hand, its use under asphalt or concrete could reduce this problem.

In relation to the contamination of water, most of the heavy metals are present as very stable and insoluble chemical compounds. Studies of leaching from slag show that the main risk of contamination of drinking water comes from lead and cadmium, but high contents of soluble chloride and sulphate also constitute a problem. Copper is particularly toxic for marine organisms [41]. Due to its potential for environmental pollution, recycling of slag calls for regulation and strict control of the amounts used, the conditions for use and possibly pretreatment so as to reduce the amount of contaminants in the slag. The identified problems highlight the need for continuous reduction in the use of heavy metals and improved sorting of the waste before incineration.

Second, organochlorines are suggested as the pollutants selected from the outdated pesticides waste stream.

References

[1] European Environment Agency (EEA), *Hazardous Waste Generation in Selected European Countries – Comparability of Classification Systems and Quantities*, Topic report No. 14/1999.

[2] OECD, *Environmental Data Compendium*, Paris, 1997.

[3] NRCs, Responses from National Reference Centers to questionnaires from European Topic Center on Waste, 1998.

[4] Junta de Residus (EPA-Catalonia), Information to the European Topic Center on Waste.

[5] National Technical University of Athens (NTUA), *Hazardous Waste Inventory in Greece, Athens*, 1999.

[6] OECD, National Accounts, Vol. II, 1997.

[7] NRCs, Comments to the European Environment Agency from National Reference Centers on Waste to draft figures for the waste chapter, July–October 1998.

[8] EUROSTAT, New Cronos Database, January 1999.

[9] ETC/W, *Baseline Projections of Selected Waste Streams*, European Topic Centre on Waste, Methodology Report, 1998.

[10] EUROSTAT, New Cronos Database, 2002.

[11] European Environment Agency (EEA), *Hazardous Waste Generation in EEA Member Countries – Comparability of Classification Systems and Quantities,* Topic report No. 14/2001.

[12] European Environment Agency (EEA), *Dangerous Substances in Waste,* Technical report No. 38, EEA, Copenhagen, 2000.

[13] European Commission, *Economic Assessment of Priorities for a European Environmental Policy Plan,* Report prepared by RIVM, EFTEC, NTUA

and IIASA for Directorate General XI (Environment, Nuclear Safety and Civil Protection), 1999.

[14] UBA, *Daten zur Umwelt – Der Zustand der Umwelt in Deutschland,* Umwelt-bundesamt, Erich Schmidt Verlag: Berlin, 1997.

[15] UNECE, *Draft Protocol to the Convention of Long-Range Transboundary Air Pollution on Persistent Organic Pollutants*, UNECE Rep. EB.AIR/ 1998/2, p. 52, 1998.

[16] Austrian Federal Waste Management Plans. *Federal Waste Management Plan, Federal Waste Management Report,* Federal Ministry of Agriculture and Forestry, Environment and Water Management, Vienna, 2001.

[17] European Environment Agency (EEA), *Europe's Environment: The Second Assessment*, Copenhagen, 1998.

[18] EEA/UNEP, *Chemicals in the European Environment: Low Doses, High Stakes?* EEA and UNEP Annual Message 2 on the State of the Environment, Copenhagen, Geneva, 1998.

[19] Lindholt, S., Unpublished report for the European Environment Agency based on CEFIC, 1998.

[20] Holland, A., O'Connor, M. and O'Neill, J., *Costing Environmental Damage: A Critical Survey of Current Theory and Practice and Recommendations for Policy Implementations,* European Parliament/STOA, Report PE 165 946/2, Luxembourg, EP/STOA, 1996.

[21] Mackay, D. *et al.*, The multimedia fate model: a vital tool for predicting the fate of chemicals. *Environ. Toxicol. Chem.*, **15**, pp. 1618–1626, 1996.

[22] OECD, Workshop on Improving the Use of Monitoring Data in the Exposure Assessment of Industrial Chemicals, Berlin, Germany, 13–15 May 1998.

[23] Van Leeuwen, J.C. *et al.*, Risk assessment and management of new and existing chemicals. *Environ. Toxicol. Pharm.*, **2**, pp. 243–249,1996.

[24] Länderausschuß für Immissionsschutz (LAI), Immissionswerte für Quecksilber/Queck-silberverbindungen – Bericht des Unter-suchungs-ausschusses, Wirkungsfragen, LAI-Schriftenreihe, Vol. 10, Erich Schmidt Verlag: Berlin, 1995.

[25] OECD, Fertilizers as sources of cadmium. *Proc. Cadmium Workshop*, 16–20 October 1995, Saltsjöbaden/Sweden, Paris, 1996.

[26] Land-Ocean Interaction Study (LOIS), LOIS Content, http://www.pml.ac.uk/ lois/index.html

[27] Wahlström, E., Hallanaro, E.-L. & Manninen, S., *The Future of the Finnish Environment*, Finnish Environment Institute: Helsinki, Edita, 1996.

[28] UNEP. Persistent Organic Pollutants: A Global Issue, http://irptc.unep.ch/ pops/

[29] UNECE. *Draft Protocol to the Convention of Long-Range Transboundary Air Pollution on Heavy Metals*, UNECE Rep. EB.AIR/1998/1, 1998.

[30] Howsam, M. & Jones, K.C., Sources of PAHs in the Environment (Chapter 4). *The Handbook of Environmental Chemistry*, Vol. 3: Anthropogenic Compounds, Part I: PAHs and Related Compounds, Chemistry, eds. A.H. Neilson & O. Hutzinger, Springer: Berlin, pp. 137–174, 1998.

[31] Keith, L.H. & Telliard, W.A., Priority pollutants: I – A perspective view. *Environ. Sci. Technol.*, **13**, pp. 416–423, 1979.
[32] Harvey, R. & Jones, K.C., Environmental chemistry of PAHs (Chapter 1). *The Handbook of Environmental Chemistry*, Vol. 3: Anthropogenic Compounds, Part I: PAHs and Related Compounds, Chemistry, eds. A.H. Neilson & O. Hutzinger, Springer: Berlin, pp. 1–54, 1998.
[33] Nolte, R.F. & Jonas, R., *Handbuch Chlorchemie I: Gesamtstofffluß und Bilanz*, UBA Texte 55/91, Umweltbundesamt: Berlin, 1992.
[34] Blomkvist, G. *et al.*, Concentrations of SDDT and PCB in seals from Swedish and Scottish waters. AMBIO, **21(8)**, pp. 539–545, 1992.
[35] Swedish Environmental Protection Agency. Pollutants, http://www.internat.naturvardsverket.se/
[36] Koss, V., *Umweltchemie*, Springer: Berlin, 1997.
[37] Thyssen, N., Pesticides in groundwater: an European overview. *Forum Book*, ed. IHOBE, 5th International HCH and Pesticides Forum, 25–27 June, 1998, Bilbao, pp. 45–54, 1999.
[38] Steurer, A., *Material Flow Accounting and Analysis*, Statistics Sweden: Stockholm, Sweden, 1996.
[39] Danish Environmental Protection Agency (DEPA), *Waste Statistics 1996*, Environmental Review no. 4, 1998.
[40] International Ash Working Group, *Municipal Solid Waste Incinerator Residues*, 1997.
[41] Thygesen, N. *et al.*, *Risikoscreening ved nyttiggørelse og deponering af slagger*, Miljøprojekt no. 203, Danish EPA: Copenhagen, 1992.

CHAPTER 2

Need and potential for underground disposal – survey of underground mines in Europe

D. Kaliampakos[1], A. Mavropoulos[2] & M. Menegaki[1]
[1]*School of Mining & Metallurgical Engineering, National Technical University of Athens, Greece.*
[2]*EPEM, Greece.*

Abstract

This chapter considers the need as well as potential for disposal of hazardous waste in underground mines and provides a comparison between surface and underground hazardous waste disposal including typical costs. A survey on underground mines in Europe is provided including some that are currently used or considered for disposal of hazardous waste in the future. The survey shows that the number of deep mines that are suitable for disposal of hazardous waste is large in Europe, especially considering that a certain number of currently used mines are expected to cease their operation in the near future. In particular, 15 EU countries are included in this survey: Austria, Belgium, Denmark, Finland, France, Germany, Greece, Ireland, Italy, Luxembourg, Portugal, Spain, Sweden, Netherlands and UK. This does not mean that the number of suitable mines in other European countries is negligible, but it only indicates that at this moment there is no sufficient data to report. It is expected that the number of suitable underground mines for hazardous waste disposal in Eastern European countries would be quite high, which offers an alternative and affordable way of dealing with hazardous waste in these countries.

2.1 Surface vs. underground hazardous waste disposal facilities

Underground hazardous waste disposal facilities present some significant advantages compared to the respective surface installations, which can be summarized as follows [1]:

- Underground facilities take advantage of the protection, isolation and security of the site. Proper design and geological siting can provide very low probabilities of hazardous substances leakage and of any such leakage to the surface environment.
- Underground structures are naturally protected from severe weather (hurricanes, tornadoes, thunderstorms, and other natural phenomena). Underground structures can also resist structural damage due to floodwaters, although special isolation provisions are necessary to prevent flooding of the structure itself. Moreover, underground structures have several intrinsic advantages in resisting earthquake motions and they tend to be less affected by surface seismic waves than surface structures [2].
- An underground hazardous waste disposal facility eliminates substantially the visual impacts, which can be of major concern in a surface structure adjacent to residential areas.
- Environmental monitoring is limited mainly to air quality within the working area. Other needs for monitoring (e.g. groundwater quality) can be determined during risk assessment.
- Long-term and after-care monitoring are usually not necessary since the main protection is provided by the geologic medium. On the contrary, in a surface hazardous waste disposal facility the protection measures have limited lifetime. Thus the landfill should be always monitored for possible leaks, even after the end of operation.
- During the operation of a surface hazardous waste disposal facility, the main cost drivers are monitoring, wastewater treatment and financial insurance. According to the above-mentioned characteristics of the underground space, operational cost is expected to be cheaper in the case of underground hazardous waste disposal.
- Moreover, in the case of an existing underground space, as it is an abandoned underground mine, there are some additional benefits that strengthen hazardous waste underground disposal, with the most important being the land cost and construction savings.

A more detailed comparison between surface and underground hazardous waste disposal facilities is given in Tables 2.1 and 2.2, while in Table 2.3 an indicative sealing cost for a surface installation, as well as the respective cost for an underground hazardous waste disposal facility are presented.

It should be noted that with the use of the techniques selected for the LowRiskDT Project, the difference between surface and underground disposal of hazardous waste would increase.

2.2 Survey of underground mines in Europe

The economic growth that has been observed in all developed European countries since the industrial revolution relied largely on mining activity. This activity

Table 2.1: General and construction issues in surface vs. underground hazardous waste disposal facilities.

	Surface HW disposal facility	Underground HW disposal facility
General issues		
Availability of space	Limited	Hundreds of abandoned underground mines may be suitable.
Sitting	Very difficult due to technical and social issues.	Easier
Licensing	Difficult	Depends on country.
Construction issues		
5 m artificial geological barrier or equivalent barrier (99/31 EC) below the waste body	Necessary	Not necessary, the use of artificial barriers is limited and it depends on risk assessment.
Leachate collection system (LCS)	Necessary, because rainfall creates huge quantities of polluted leachate. Normally LCS constitutes of extended piping and drainage layer.	Not necessary if water does not enter the waste body. A kind of LCS should be constructed for potential leaks.
Wastewater treatment	Necessary. Treatment level depends on local conditions and potential impacts at water tables and most of the times should be a third level one.	Most of the time, negligible or no wastewater is generated. Safe storage of wastewater and transfer to wastewater treatment facilities is an indicated solution.
Storm water management	Necessary, one of the basic components of design and construction.	Depends on the underground mine conditions – may also be negligible.

Table 2.2: Operational and cost issues in surface vs. underground hazardous waste disposal facilities.

	Surface HW disposal facility	Underground HW disposal facility
Operational issues		
Stability	Crucial point for the waste body formulation.	Crucial point for the underground space.
Environmental impacts of possible major accidents (SEVESO)	High impacts to water and ground/soil. Toxic gases emissions are considered as a high level hazard.	Limited or no impacts to water and ground/soil system. Toxic gases emissions may create problems to workers.
Environmental monitoring	Extended monitoring is necessary, especially for water and air quality. The sensitivity of the surrounding ecosystem and natural resources determines more specific areas that should be monitored.	Monitoring is limited to air quality, within the working area. Risk assessment determines other needs for monitoring.
In situ treatment options	Easier	More difficult due to space limitations.
Long-term–after-care monitoring	All the protection measures have limited life-time, thus the landfill should be always monitored for possible leaks, even after the end of operation.	The main protection is provided by the use of underground space – the deeper the better. After-care monitoring is not necessary.
Cost issues		
Construction cost	Artificial barriers, wastewater treatment, leachate collection system, gas collection and treatment system, storm water management and excavations are the main components.	The components may be the same, but they will probably be cheaper due to limited water entry and utilization of already available space.
Operational cost	Monitoring, wastewater treatment and financial insurance, are the main cost drivers.	It is expected to be cheaper.

Table 2.3: Typical sealing cost in surface and underground hazardous waste disposal facilities.

	Thickness (m)	Quantity	Unit	Cost per unit (euros)	Cost per m² (euros)
Bottom layer for surface installations					
Clay barrier (hydraulic conductivity $< 10^{-9}$ m/s)	5	8.5	m^3	10	85
Geotextile	–	1	m^2	2	2
HDPE geomembrane (hydraulic conductivity $< 10^{-9}$ m/s)	0.002	1	m^2	6	6
Geotextile	–	1	m^2	2	2
Drainage layer	0.5	0.5	m^3	5	2.5
Total cost					97.5
Sealing for underground facilities					
HDPE geomembrane	0.002	1	m^2	15	15
Shotcrete	0.1	1	m^2	32	32
Total cost					47

was reflected in a large number of mining exploitations, many of which were underground mines. However, the decline of the mining industry during the last decades has led to the closure of many mining sites throughout most European countries. As a result, there are many abandoned underground mines which most of the times remain inactive and practically useless. In addition, due to the continuous decline of the mining industry, a large proportion of the remaining underground mines are expected to cease their operation in the near future [3, 4]. These mines could also be considered as potential disposal sites.

A profile of the mining activity has been formulated for all the 15 EU countries [5]. The profile consists of some general data concerning mining activities, active mines and mineral production, active and inactive mines etc. Special emphasis has been given to identify the underground mines in order to look for more details. More than 70 underground mines were registered and their main characteristics were recorded. Most mines are located in Germany, Sweden, Finland, and the UK, as expected due to the intense mining activity in these countries (Fig. 2.1).

In addition, an inventory of inactive underground mines, presently used as waste disposal sites, has been carried out.

Both of the previous results are presented hereinafter.

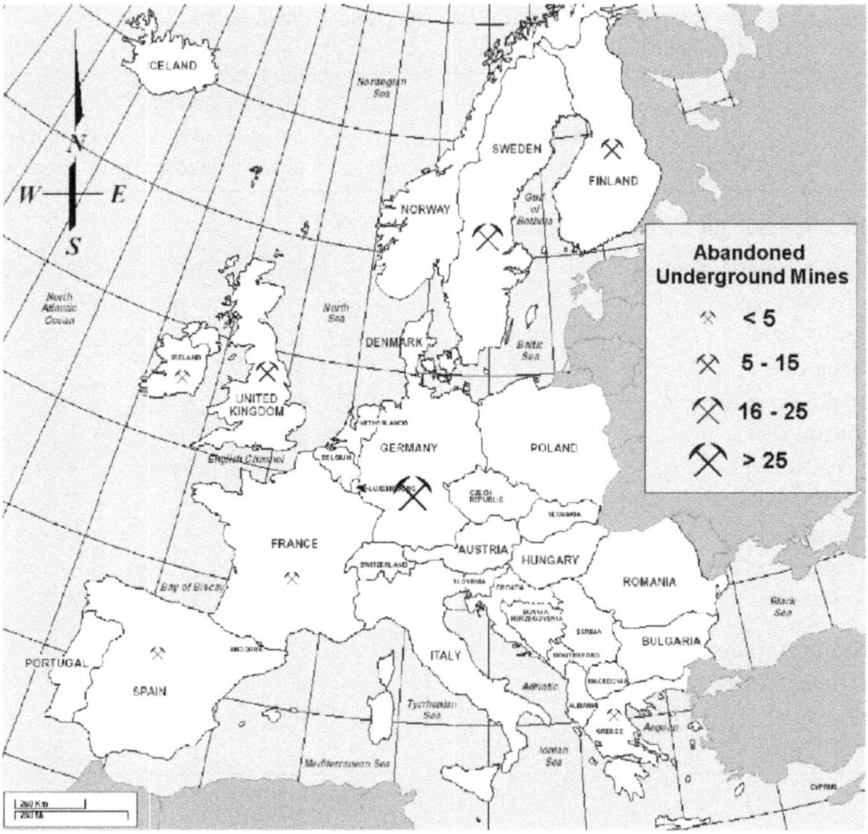

Figure 2.1: Distribution of abandoned underground mines.

2.3 The profile of mining activity in 15 EU countries

2.3.1 Austria

Although the mining industry has maintained a long tradition in Austria, the metal mining sector is declining, principally due to high operating costs, low ore grades, environmental problems and increased foreign competition. This is not the case with the industrial minerals sector, which produces a number of important minerals. Austria is considered to be a significant world producer of graphite, magnesite and talc [6].

Because of Austria's long history of minerals exploration and mining tradition, geologic conditions are fairly well known. Future mining activities will most likely be concentrated in industrial minerals, mainly for domestic consumption. The chances of finding new and workable base metal deposits are probably remote.

No specific information could be retrieved on inactive mines of the country except Schmitzbe coal mine, which closed in 1995, and Trimmelkam, which closed in 1992.

2.3.2 Belgium

Although Belgium has a significant mineral-processing industry, it does not produce minerals as a result of mining activities. In fact, Belgium has no economically exploitable reserves of metal ores or primary energy.

Belgium has a significant industrial minerals sector and is an important producer of four groups of industrial materials: carbonates, including limestone, dolomite, and whiting; synthetic materials in the form of soda ash and sodium sulphate; silica sand; and construction materials, including a wide range of different types of marble [6].

Following the closure of the last coal mines in 1992, the only mining operations left in Belgium in 1998 were for the production of sand and gravel and the quarrying of stone, principally specialty marbles and the Belgian blue-grey limestone called 'petit granit'.

Very little information has been retrieved about inactive mines in Belgium. The only abandoned mines found are some coal mines located throughout the country.

2.3.3 Denmark

Denmark's mineral resources are, mainly, the natural gas and petroleum fields in the North Sea that, together with renewable energy, have made Denmark a net exporter of energy since 1996. Most of the mineral commodities produced in Denmark are exported with the majority shipped to EU countries.

The mining and metal industry works closely with the Ministry of Environment and Energy, the Danish Environmental Protection Agency, local and community governments, and citizen groups to minimize any adverse effects to the environment. Environmental protection is the main focus of the Danish Environmental Protection Agency. A common goal of the steelworks and other industrial concerns is to make use of as much raw material taken into the plant as possible and to maximize the use of any by-products, such as flue dusts.

Denmark has large reserves of non-metallic materials such as chalk, diatomaceous earth, limestone, and sand and gravel. Approximately one-third of the bedrock area in Denmark consists of chalk and limestone. Denmark's industrial minerals sector is based mainly on these easily accessible materials. Cement, chalk for paper filler, ground limestone and lime, including agricultural and burnt, are produced [7].

As far as sand, gravel and aggregates are concerned, from the mid-1980s to the mid-1990s, the industry was suffering from low prices and fierce competition. However, due to the upswing in the Danish building and construction industry, the industry is now in a healthier shape.

Denmark is the only commercial producer of moler, which consists of a natural mixture of diatomite and 20–25% bentonite. Moler has a variety of applications, such as industrial absorbers, brake linings, and fertilizers, and is an important ingredient of insulation bricks [6].

No specific information could be retrieved on the inactive mines of the country.

2.3.4 Finland

Mining history in Finland dates back to 1540, when the quarrying of iron ore commenced in the southern part of the country. Since then, about 260 metallic mines have been operated, with the total amount of ore extracted being around 250 Mt. Finnish metallurgical technology and manufacturers of mining equipment are well-known throughout the international mining community. The exploitation of copper, nickel, cobalt, zinc and lead ores as well as chromium, vanadium and iron deposits has provided the raw material base for the country's metal industry, with significant processing and refining of copper and nickel concentrates at Harjavalta, zinc at Kokkola, chromium at Kemi by the Outokumpu Group and iron at Raahe by Rautaruukki Oy. The major industrial minerals mined in Finland are apatite, talc and, to a lesser extent, limestone [8]. On 1 January 1995, Finland acceded to the EU. At that time, amendments to the Finnish Mining Law concerning reciprocity took effect and allowed any individual corporation or foundation having its principal place of business or central administration within the EU, to enjoy the same rights to explore for and exploit mineral deposits as any Finnish citizen or corporation. This encouraged foreign investment and increased exploration activities of major and junior companies. Exploration emphasis was given on base metals, diamond, and gold deposits.

There are many inactive mines in Finland. Data are included in the websites of Geological Survey of Finland and Outokumpu Oy, which is the leading company in the country. However, due to the lack of available data it could not be specified whether they are open-pit or underground mines.

2.3.5 France

France is a major European mineral producer. The traditional mineral industries have been in a state of transition a few years ago. In the past, the heavy economic and political involvement of the state was one of the main elements of the national mineral policy. During the last years, efforts have been made to promote the private sector and to reduce the dependence of state-owned companies on subsidies.

The government proceeded with a programme of privatization involving large state-controlled companies to reduce the direct role of the Government in the economy. Among the nine major companies privatized since 1994, the Péchiney Group, Rhône-Poulenc S.A., Société Nationale Elf Aquitaine, and the Usinor Group were included [6].

Several industries, such as bauxite, coal, iron ore, and uranium, have steadily undergone changes during the past few years, especially bauxite, which is no longer mined.

The iron ore basin of northern France stretches from Lorraine northward into Belgium. For many years, the high phosphorus and low iron content of the ore limited its desirability and the production has been declining for several years. Terres Rouges Mine, the last iron ore mine in the Lorraine district, closed in 1998. French bauxite production ceased altogether by the end of 1993. Mining of lead and zinc completely ceased in France. The two working potash mines, Amelie and Staffelfelden, will be closed until the end of 2004 [9]. All underground coal mines were closed in the Midi-Pyrénées region and in the Nord Pas-de-Calais Basin. Mining in La Mure (Isére) and Carmaux (Tarn) ceased in 1997 [6]. Charbonnages de France envisioned the final stoppage of all coal mining in France by 2005.

2.3.6 Germany

The minerals and metals industry, which includes industrial processing, construction, and the mining industry, contributes almost 1% to the GDP. Production in the mining and metals industries depends on a variety of forces, including the availability of materials, as well as the supply and demand. The easing of the worldwide recession is a positive factor for those industries that depend on the exportation of their products. The high costs of production in Germany compared with those of competing foreign producers and the problems caused by trying to balance production between the merged German Democratic Republic and Federal Republic of Germany led to constraint production [6].

The technological standard of German mining operations is world class. Notwithstanding the general contraction of the industry, the production levels of certain minerals remain important, both domestically and on a global scale. For example, lignite ranks 1st in the EU and in the world; marketable rock salt and potash, 1st in the EU and 3rd in the world; and hard coal, 1st in the EU and 11th in the world.

There are a large number of inactive mines located in Germany. It should be specified though that there is not much information about their present condition.

2.3.7 Greece

The mining and metal-processing sectors of the economy of Greece are small but important parts of the national economy. The mining sector's share of the gross national product is 1.7%. They are highly concentrated, as five mining companies handle approximately 60% of the sector's turnover. Bauxite is the most important of the Greek mineral commodities. Other important commodities are chromite, gold, iron, lead, nickel, and zinc [6]. Greece is the largest producer of bauxite and nickel in the EU.

Northern Greece is thought to contain a significant amount of exploitable mineral resources and is receiving more attention with regard to exploration activities.

In 1998, most activities were directed toward gold. A number of multinational companies, such as Rio Tinto Plc., Normandy Mining Ltd., and Newmont Inc., expressed their interests in Greece's northern territories [10].

The Kassandra Mines (Skouries and Olympias deposits) in northern Greece have been producing lead, silver, and zinc for more than 30 years. The mines were bought in 1996 by TVX Hellas, an affiliate of TVX Gold Inc. of Canada, with the idea of exploiting the refractory gold ore by incorporating pressure oxidation technology into the ore-processing phase.

To date, the extracted ore could not be processed, due to the opposition of the residents from nearby areas, who were against the operation of a processing plant due to environmental problems.

General Mining & Metallurgical Co. S.A. a ferronickel producer, was the latest state-owned company to be put up for sale by the Greek Government. LARCO was one of the world's highest cost producers of nickel in ferronickel.

Greece is the world's second-largest producer of bentonite after the USA. Bentonite is extracted from the island of Milos by open-pit mining. S&B and Mykobar Mining Co. S.A. (acquired by S&B in March 1999) are the major producers and accounted for almost all of the Greek bentonite.

S&B, together with its affiliates, is the largest producer of perlite in the EU. Perlite is extracted from the island of Milos by Otavi Minen Hellas S.A. (purchased by S&B in 1998). S&B continued also the production of natural zeolite in northern Greece.

Lava Mining and Quarrying Co. S.A. (LARCO), specializes in industrial minerals with production of gypsum from the island of Crete, pozzolan from Milos, and pumice from the island of Yali.

Grecian Magnesite S.A. is a leading magnesite producer in the western world and the biggest exporter in the EU. Its open-pit mine is at Yerakini in northern Greece.

The Greek marble industry plays a leading role in the international dimension stone market, as a result of the marble production in almost all areas of the country, its variety of uses and many colours (ash, black, brown, green, pink, red, and multicoloured).

PPC is the major producer of lignite, the predominant fuel in electricity generation in Greece. PPC continued exploration in the basins of Amyntaion, Elasson, Florina, Megalopolis, and Ptolemais. PPC had reserves estimated to be 6.8 billion tons from which 4 billion tons was estimated to be economically recoverable by open-pit mining. Most PPC lignite is produced from the Ptolemais-Amyntaion basin with lesser amounts from the Megalopolis basin.

There are various inactive mines in Greece among which there are four underground mines.

2.3.8 Ireland

The exploitation of minerals in Ireland has a long history with small-scale production up to 1969. In that year the large complex lead–silver–zinc–copper–barite Tynagh deposit was discovered and several others followed [11].

Ireland is a major EU producer of zinc and an important producer of alumina, lead, and peat. Although the range of minerals exploited in the country has been limited, exploration activity for new mineral resources is continually increasing, mainly emphasized in gold, lead, and zinc. The country's mineral-processing industry is small, as is the demand and consumption of mineral products [6].

Today, there are only three active mines in Ireland: the Tara mine, the Galmoy and Lisheen mine.

Industrial mineral production in Ireland is rather low with gypsum and limestone (production of about 1 million tons) being the most important.

There are four inactive mines in Ireland, three of which are underground. There are also two inactive mines, but no information was found on whether they are underground or open-pit.

2.3.9 Italy

Italy is a significant processor of imported raw materials, as well as a significant consumer and exporter of mineral and metal semi-manufactured and finished products. It is the world's largest producer of pumice and related materials, producing almost one-half of the world's output, as well as the world's largest feldspar producer, producing about one-fourth of the world's output. The country is the world's eighth and tenth largest producer of crude steel and cement, respectively. Italy is also an important producer of dimension stone and marble [6].

Growth in Italy's mining and extractive industries was marginal in 1998. Among the metallic ores, lead was mined, although production was minimal and decreasing. Most of the output came from the Silius Mine in Sardinia. The small output of zinc ore came from the safety and environmental recovery work in the remaining sites in the Iglesias area of Sardinia.

Industrial mineral production is the most important sector.

Italy is the second largest cement producer of the EU, after Germany. Italcementi-Fabbriche Riunite Cemento S.p.A. is the largest cement producer in Italy with 28 plants and more than 30% of the Italian market.

Italy is famous for its marble, which occurs in many localities and is quarried by hundreds of different companies.

In 1998, production of potash remained suspended. The main reasons were the result of a severe drought that has restricted the availability of process water to the plants and the inability to remove waste material and mine water owing to environmental and ecological concerns. In Sicily, the underground mines that were operating at Pasquasia, Racalmuto, and Realmonte, remained on standby.

Mining of metallic ores is expected to remain at its reduced level because of ore depletion. The metals-processing industry, based primarily on imported stocks, is expected to continue to play an important role in Italy's economy. Italy is expected to remain a large producer of crude steel and a significant producer of secondary aluminium in the EU.

The industrial minerals quarrying industry and preparation plants are expected to remain significant, especially in the production of barite, cement, clays, fluorspar, marble, and talc. Italy is expected to continue to be the world's leading producer of feldspar, feldspathic minerals, and pumice. The ceramics sector is expected to be important, particularly regarding exports.

2.3.10 Luxembourg

Luxembourg's mineral industry consists principally of raw materials processing information systems, and trading, among others. The country produces traditionally sand and gravel and crushed and dimension stone.

Mining in Luxembourg is represented by small industrial mineral operations that produce material for domestic consumption. These minerals include dolomite, limestone, sand and gravel, and slate [6].

ARBED dominates the mineral industry and is involved in producing pig iron, crude steel, and stainless steel, all from imported material. The company specializes in the production of large architectural steel beams and is involved in other areas of the economy, such as the cement and brick-making industries. ARBED's domestic and foreign subsidiaries have interests in steel making and steel products, cement, copper foil production, engineering, and mining.

As stated above, mining activity in Luxembourg is very limited and consists of domestic-scale industrial minerals operations. Thus, no specific information could be retrieved on active and inactive mines of the country.

2.3.11 Portugal

Portugal has a long history in the exploitation of metallic minerals starting about 2000 B.C. [6]. The first mining operations took place in 'gossan' type oxidation zones (for copper, zinc, lead, gold and silver) and gold-bearing placers. Later, the Romans intensively exploited gold and polymetallic sulphide vein deposits [12].

From a geological point of view, Portugal is a considerably diverse and complex country. More specifically, the Iberian Peninsula is one of the most mineralized areas of Western Europe with a very complex geology. Massive sulphides linked to synorogenic vulcanism in the southwestern part of the Iberian Peninsula are well known internationally. The metallogenic province stretches about 250 km from Seville, Spain, to the southwestern coast of Portugal. On that world famous district a total of 30 deposits (11 in Portugal and 19 in Spain), with more than 1120 Mt, were discovered between 1950 and 1998, averaging 1.2 deposits/2 years, which is an amazing exploration performance index.

Today, Portugal is a significant European mineral producer and one of Europe's leading copper producers. It is also a major producer of tin, tungsten, uranium and marble.

The Neves-Corvo Mine owned by Somincor and Rio Tinto Ltd. and the Panasqueira tungsten mine of Beralt Tin and Wolfram (Portugal) Ltd. are the two major operations in the metal-mining sector.

In Portugal there is no current gold production. However, a number of deposits have been identified and considered to be significant. Jales-Tres Minas is the most important gold district in Portugal while Auspex Minerals Ltd., also announced in 1998 that they discovered 13 deposits with potential economic gold mineralization.

Industrial minerals production in Portugal is represented by a variety of materials, most notably ceramics and dimension stone. The dimension stone industry is an important segment of the mining industry in terms of value and trade. Marble is the most valuable of the stone products and accounts for the majority of stone production. The main area for marble mining is the District of Evora. There is a potential for increased production of granite, marble, and slate. In addition, Pirites Alentejanas S.A.R.L. is the country's largest producer of pyrite.

The present structure of the mineral industry could change in the near future because of significant mining exploration by several foreign companies. Copper, gold, kaolin, lead, lithium, pyrites, and tin are some of the minerals targeted for exploration. The Iberian Pyrite Belt is the prime area for exploration activity and appears to have an above-average potential for success on the basis of district's record of about 90 documented mineralized deposits, an unusually high number of large sulphide deposits.

According to the Geological and Mining Institute of Portugal, there are numerous inactive mines in Portugal.

2.3.12 Spain

Spain is a significant European producer of non-ferrous precious metals, with some of the most mineralized territories in Western Europe. The main polymetallic deposits, from west to east, include Tharsis, Scotiel, Rio Tinto, and Aznalcollar. There are very few large mines. In terms of value of metallic and non-metallic minerals and quarry products, Spain is a leader among the EU countries. Consequently, Spain has one of the highest levels of self-sufficiency, with respect to mineral raw materials, among the EU members. Of a total of approximately 100 mineral products mined, about 18 are produced in significant quantities, such as bentonite, calcinated magnetite, copper, fluorspar, glauberite, iron, lead, mercury, potassic and sepiolitic salts, pyrites, quartz, refractory argillite, sea and rock salt, tin, tungsten, and zinc [6].

Production of many metallic minerals in Spain is insufficient to meet domestic demand, so these must be imported. For most non-metallic minerals, however, production exceeds by far domestic consumption and the surpluses are exported. The economic development of certain regions, such as the Basque Country and Asturias, is based on their mineral wealth. Therefore, mining is an important current and potential source of income in these areas.

Spain is one of the larger coal producers in the EU, with 26 million metric tons per year (Mt/yr) (all types), in 1998. Coal reserves are abundant but difficult to mine. Consequently, cost of production is higher, making Spanish coal less competitive than that of many other countries. The leading producer of soft coal is Huelleras del Norte S.A. (Hunosa) and the leading producer of lignite is Empresa Nacional de Electricidad S.A. (Endesa).

Copper is mainly mined at the deposits in Sotiel and Migollas in Huelva, by Navan Resources Ltd. (Almagrera) and by Boliden Apirsa at Aznalcollar (Los Frailes deposit) near Seville.

Gold was being sought in Asturia, northeastern Spain, by Rio Narcea Gold Mines, Ltd., which acquired concessions and permits that previously belonged to the Spanish subsidiary of Anglo-American Corp.

Navan Resources Ltd. inaugurated its new polymetallic (copper, lead, and zinc) Aguas Tenidas Mine near Huelva in November 1997. Aguas Tenidas is the first underground operation to be developed in Spain in several years. The operation supplies Navan's nearby Almagrera mill and concentrator with 0.8 to 1 Mt/yr of ore. Navan acquired the mill and concentrator, along with three mines, Sotiel, Sotiel Este, and Miggollas, in June 1997.

The principal producer of iron ore was Compania Andaluza de Minas S.A. (CAM), which operated its open-pit mine at Marzuesado (Granada). Mining was halted in October 1996, and the mine remains inactive since the end of 1997.

However, production started at the nearby Los Frailes, one of the biggest open-pit mines in Europe. Ore production at Los Frailes was estimated to be approximately 4 Mt/yr. Los Frailes was closed in early 1998 after a large toxic spill. A waste reservoir ruptured and sent sludge into a nearby river. The spill poisoned some of the areas around the edges of Donana National Park, Europe's largest nature reserve. Boliden was undertaking remedial actions and safety requirements in order to restart operations as soon as possible.

There are a number of inactive mines in Spain. No specific information could be retrieved on the inactive mines of the country, except that most of them are coal mines.

2.3.13 Sweden

Sweden is endowed with significant deposits of iron ore, certain base metals (copper, lead, and zinc) and several industrial minerals, including dolomite, feldspar, granite, ilmenite, kaolin, limestone, marble, quartz and wollastonite. The country is well known for the production of high-quality steel. Sweden has developed nuclear and hydroelectric power, since the country must rely heavily on hydrocarbon imports owing to inadequate indigenous resources. After acceding to the EU on 1 January 1995, Sweden liberalized its mineral policy to parallel EU standards. The policy, based on the Swedish Minerals Act, 1992, eliminated laws requiring foreign companies to get special permission for prospecting, annulled the state's participation in mining enterprises (so-called 'crown shares') and revoked all taxes and royalties, except for a 28% corporate tax, one of the lowest in Europe. Furthermore, an exploration permit holder cannot receive an exploration permit until adequate financial and technical capabilities can be proven [6].

The two largest companies in Sweden are Boliden AB, owned by Boliden Ltd, and the government owned Luossavaara-Kiirunavaara AB (LKAB).

Laisvall is the largest lead mine in Europe and it is located in Arjeplog Municipality, in northern Sweden, towards the Norwegian border. In January 1999, total proven and probable reserves were 6.8 Mt grading 0.8% zinc, 4.6% lead and 11 g/t silver. Measured and indicated reserves at that time were 3.35 Mt grading 1.2% zinc, 2.0% lead and 9 g/t silver. The company has planned to increase the ore output rate from Laisvall to 2.2 Mt/yr, given regulatory approvals [13].

Located near Hedemora, in the historic Bergslagen mining district of central Sweden, the two mines and common concentrator at Garpenberg comprise the smallest of Boliden's mining areas. The company bought the Garpenberg mine and mill from AB Zinkgruvor in 1957. The exploration of a silver-rich area to the north during the 1960s led to the development of a second mine, Garpenberg Norra (Garpenberg North).

The open-pit Bjorgdal mine is the largest gold mine in Western Europe. The former owner, Terra Mining AB, was bought by Williams Resources Inc. in 1996. Williams Resources was continuing exploration activities and reported in 1998 that it had increased estimated minable reserves to 8.6 Mt of ore grading an average of 2.32 g/metric ton gold [14].

LKAB has iron ore mines and processing plants in Kiruna and Malmberget, a pelletizing plant in Svappavaara, and harbors at Luleå and Narvik. The company operated close to full capacity in 1997. LKAB's Malmberget (ore mountain) iron ore mine, located at Gällivare, 75 km from Kiruna, contains some 20 orebodies spread over an underground area of about 5 by 2.5 km. Seven are currently being exploited. Mining began in 1892 and since then over 350 Mt of ore have been produced. Kiruna has the world's largest underground iron ore mine. The orebody in Kiruna is an enormous slice of magnetite. It is about four kilometres long, has an average width of 80 m and extends to an estimated depth of around 2 km at an incline of roughly 60°. The main haulage level is at a depth of 1.045 m. Mining of the orebody between levels 1.045 and 775 will continue until about the year 2018. Up to now, about 940 million tons of ore have been extracted from the Kiruna orebody.

The Zinkgruvan Mine, the largest zinc mine in Sweden, is owned by North Mining Svenska AB, a subsidiary of the Australian company, North Limited. Underground mining started in 1857. In the early 1990s, new technology and careful management reduced mining and milling costs to about 50%, converting a high-cost operation to the sixth lowest-cost zinc producer in the Western World by 1993. Currently, the operation is producing about 700,000 t/yr of zinc in concentrate.

The total production of industrial minerals, except aggregates and dimensional stones, in 1997 reached 9 million tons, a level that has been fairly constant during the 1990s. Limestone products, including dolomite and limestone for cement production, form 90% of the total, while silica sands, quartzite, feldspar, olivine and talc make up for the remaining 10% of the output. Tricorona Mineral AB owns three major mineral deposits, namely graphite, kaolin and wollastonite, of which only the graphite was in production in 1998. Three subsidiaries were formed to handle the development of the deposits, Woxna Graphite AB, Svenska Kaolin AB and Aros Mineral AB respectively [15].

According to the Swedish authorities on underground exploitation, the total number of abandoned mines is 25 in Northern Sweden and 775 in Central and Southern Sweden.

2.3.14 The Netherlands

In terms of world production, the Netherlands is a modest producer of metallic and non-metallic minerals and mineral products. Production of mineral commodities generally remained the same or dropped slightly in 1998, compared to previous years. The high cost of social benefits contributed to the production costs of Dutch products making them less competitive on the world market. The only mining operations left in the Netherlands are the extraction of peat, salt, and sand and gravel. The metal-processing sector relies almost exclusively on imported raw materials [6].

The Netherlands has no commercially exploitable reserves of metal ores. The only active mines that exist in the country extract industrial minerals.

No specific information could be retrieved on inactive mines of the country.

2.3.15 The United Kingdom

Mine production of ferrous and non-ferrous metals in the UK has been declining for the past 20 years as reserves become depleted. Since processing is the basis of a large and economically important mineral industry, significant imports are required to satisfy metallurgical requirements [6].

Operations in the steel sector showed moderate increases as the demand for steel increased. The industrial minerals sector has provided a significant base for expanding the extractive industries, and the balance has shifted away from the metallic mineral sector. Companies had a substantial interest in the production of domestic and foreign industrial minerals, such as aggregates, ball clay, gypsum, and kaolin (china clay).

Production of iron ore is limited to a small amount of hematite ore, mined by Egremont Mining Co. at the Florence Mine in Cumbria. The output goes for pigments and foundry annealing uses, rather than metal production. Primary steel production is based on imported iron ore, mainly from Australia and Brazil.

Activities in gold exploration and development in the UK increased in 1998. Northern Ireland, Scotland, and Wales continued to be the three main areas of exploration by companies. Scotland was the most active area with several exploration licenses in effect.

The UK is the leading world producer and exporter of ball clay, as well as the world's largest exporter and second largest producer, after USA, of kaolin (china clay).

Watts, Blake, Bearne & Co. Plc. (WBB) is the country's largest producer of ball clay. WBB Devon Clays Ltd. is responsible for the ball clay operations of WBB. The division operates eight open-pits and three underground mines that have a total combined capacity of 500,000 t/yr of crude ball clay.

English China Clays Plc. (ECC) is the largest producer of kaolin and one of the major producers worldwide. Operations are mainly found in the southwestern area of the UK. ECC Ball Clays Ltd. is responsible for the domestic ball clay operations of ECC. The division operates five quarries and three underground mines that have a combined output of 450,000 t/yr of crude ball clay. ECC International Ltd. operates ball clay and kaolin mines and quarries in the Wareham Basin, Dorsetshire; the Bovey Basin, South Devonshire; and the Petrockstowe Basin, North Devonshire. The majority of the production comes from the Bovey Basin.

Fluorspar mining is concentrated in Derbyshire, from the Southern Pennine deposit. The major producer is Laporte Industries Plc., which operates two underground mines and one open-pit mine. The ore is processed at Laporte's Cavendish Mill near Sheffield.

Durham Industrial Minerals Ltd. was to close five fluorspar mines at Rookhope in Weardale. Falling prices of fluorspar, Chinese competition, and the strength of the pound were thought to have contributed to the closings [16].

British Gypsum Ltd., a subsidiary of BPB Industries Plc., is the major producer of gypsum in the UK. The company has mines in Cumbria, Leicestershire, Nottinghamshire, Staffordshire and Sussex that produce about 3 Mt/yr of gypsum. With few exceptions, this material supplies the domestic market.

Cleveland Potash Ltd. (CPL), the only potash producer in the UK, operates the Boulby Mine in Yorkshire. CPL also mines rock salt as a co-product from an underlying seam in the Boulby Mine. Boulby potash occurs at depths between 1200 and 1500 m in a seam ranging from 0 to 20 m but averaging 7 m in thickness [17].

Most slate mining in the UK occurs in northern Wales; additional mining operations are found in Cornwall and the Lake District. Alfred McAlpine Slate Ltd. is the owner and operator of the Cwt y Bugail, Ffestiniog, and Penrhyn quarries in North Wales. The Penrhyn quarry at Bethesda, measuring 2.415 by 805 m, is considered to be the world's largest slate quarry and has been in operation for more than 400 years. The company also produces natural slate from its American quarry at Hilltop Slate Inc., New York. Historically, natural slate has been used in roofing applications, but in more recent times, markets have been extended to include interior flooring and windowsills together with ornamental landscapes. McAlpine Slate produces more than one-half of the UK's entire production of natural slate. The company exports about two-thirds of its production, mostly to Europe. McAlpine received planning permission to exploit additional reserves at its Penrhyn quarry. The quarry, which covers an area of about 325 hectares (h), will be extended by an additional 45 h. This enlargement will extend the life of the quarry and increase extraction by a further 80 million metric tons of slate at the southern end of the quarry [18].

RJB Mining Plc., the largest coal mining company in the UK and the largest independent coal producer in the EU owns most of the coal mining industry. The largest operation is the underground Selby Complex, consisting of Riccall/Whitmoor, Stillingfleet Combine and Wistow. There were also 24 small drift mines in operation in 1998. Open-pit mines in production in 1998 totalled 83. RJB Mining owned 16 producing open-pit mines; Celtic Energy Ltd. owned 5 open-pit mines; and Scottish Coal Company Ltd. had 11 open-pit mines in Scotland. The remaining open-pit mines were operated by more than 25 other operators.

The UK has been a significant player in the world mining and mineral-processing industries. This has been more the result of an extensive range of companies in the country, with various interests in the international mineral industry rather than the domestic mineral industry. This scenario is expected to continue.

Exploration is expected to continue onshore and offshore. Onshore exploration activities will be directed mainly toward precious metals. Offshore exploration interest will continue to be focused on North Sea areas, particularly the areas west of the Shetland Islands, the Central North Sea, and the Southern Gas.

Five large underground mines in the UK ceased operations in the period 1998–2000.

2.4 Inactive underground mines used as waste disposal sites

Although the storage of wastes in inactive underground mines has attracted considerable interest in the past twenty years, it could be considered as a fairly recent concept. Salt mines, which usually are excavated by the room and pillar method, are of great interest in view of the possibility of reusing the openings for waste disposal. Some examples of inactive underground mines that have been used as waste repositories are shown in Table 2.4.

Several studies have been conducted on the feasibility of a deep geological disposal site and various geological media have been analysed for their thermal, mechanical and chemical properties. As a result, four underground research laboratories are currently in operation in Europe: crystalline granite is being investigated at Grimsel (Switzerland) and Stripa (Sweden); the suitability of clay analysed at Mol (Belgium) and a salt formation is being studied at Asse (Germany). Furthermore, laboratories are scheduled for the near future or are already under construction, namely in France, Sweden (Aspo) and the UK (Sellafield). It must be specified that both Stripa and Asse are inactive underground mines [20]. Major past or present underground research laboratories are shown in Table 2.5.

The information below is a brief description about underground inactive mines that have been used as waste repositories, underground laboratories and for research purposes related with storage of wastes.

2.4.1 Morsleben salt mine

Morsleben repository is located in the federal state of Saxony-Anhalt [22]. At the site, potassium was mined until the early twenties. Thereafter, rock salt mining went on until 1969. Both the above operations left open cavities with a volume of approximately 10 million m^3.

In 1970 the nuclear power plant operator of the former German Democratic Republic bought the mine to convert it into a low-level (LLW) and intermediate-level waste (ILW) repository. After a licensing procedure, waste disposal started in 1978 using rock cavities below the 500 m horizon for waste emplacement.

Table 2.4: Examples of inactive underground mines that have been reused as waste repositories in Europe [19].

	Name of mine	Type of ore	Type of reuse	Notes
Germany	Bartensleben mine	Salt	Storage of radioactive wastes	
	Konrad mine	Iron	Storage of radioactive wastes (under study)	
	Heilbroun mine	Salt	Storage of fly ash wastes; storage of anhydrite and clay contaminated with Hg	Room and pillar
	Kochendorf salt mine	Salt	Storage of flue gas, desulphurization residue from incineration plants and siliceous slags (under study)	Room and pillar
	Walsum mine	Coal	Storage of fly ash from incineration plants in the goaf (under study)	Longwall mining
	Haus Aden/Monopol mine	Coal	Storage of fly ash from incineration plants in the goaf (under study)	Longwall mining
	Zielitz mine	Potash	Storage of industrial wastes	Room and pillar
	Morsleben mine	Salt	Storage of radioactive wastes and sealed radiation sources	Room and pillar
	Herfa-Neurode mine	Salt	Storage of hazardous wastes	
France	Joseph-Else mine	Potash	Storage of industrial wastes (under study)	Room and pillar
Italy	Codana mine	Gypsum	Storage of industrial wastes	Room and pillar
	Besta mine	Dolomite	Storage of inert debris (36.000 m^3 reused)	Room and pillar
Russia	Verkhnekamsoye area mines	Potash	Storage of waste	Room and pillar
Slovenia	Velenje mine	Coal	Storage of fly ash (under construction)	Longwall mining
UK	Walsall Wood colliery old mine	Coal	Storage of chemical wastes (since 1965)	
	Dudley mines	Limestone	Colliery waste and fly ash pumped in the voids from the surface	Room and pillar
	Geostow project	Salt	Storage of fly ash from incineration plants	

Table 2.5: Major past or present underground research laboratories [21].

Rock formation		Laboratory name	Country
Salt	(bedded)	Salt Vault (Kansas)	USA
	(dome)	Avery Island (Louisiana)	USA
	(dome)	Asse	Germany
	(bedded)	WIPP (New Mexico)	USA
	(dome)	Hope	Germany
Crystalline rock	(granite)	Stripa	Sweden
	(granite)	Grimsel	Switzerland
	(granite)	Edgar mine (Colorado)	USA
	(granite)	Tono mine	Japan
	(granite)	URL (Manitoba)	Canada
	(granite)	Climax mine (Nevada)	USA
	(granite)	Fanay Augeres	France
	(granite)	Akenobe mine	Japan
	(granite)	Hard Rock Laboratory	Sweden
	(granite)	NSTF (Washington)	USA
	(basalt)	G-tunnel (Nevada)	USA
Argillaceous rock	(plastic clay)	Mol	Belgium
	(clay-marl)	Pasquasia	Italy

Morsleben became a Federal Facility following German reunification, DBE was then contracted to operate the site.

In this deep geological repository different categories of solid LLW and ILW as well as sealed radiation sources are disposed of. Essentially, LLW packed in drums is stacked in chambers, while waste with higher activity content, delivered to the repository in shielding overpacks, is discharged through shielding lock systems into closed chambers below a drift. Waste disposal (Fig. 2.2) is carried out on the basis of contractual arrangements between waste producers and the Federal Government. Ownership of the waste is passed over upon delivery; the producers pay a fee that settles for all costs.

In 1998, the radioactive waste disposed at Morsleben amounted to 36.752 m^3 radioactive waste and 6.621 sealed radiation sources.

2.4.2 Herfa-Neurode salt mine

The Herfa-Neurode underground waste repository (Figs 2.3–2.5) is owned by Kali und Salz Entsorgung GmbH, which also operates another underground waste repository named Zielilz [23]. Hazardous waste disposal has been undertaken there for the last 30 years. The underground waste disposal plant is located in a mining concession of the potash mine Winterschall at Heringen/Were in Germany. The mine is situated in a 300 m thick salt formation, covered by clay

Figure 2.2: Waste disposal at Morsleben.

Figure 2.3: Surface view of the Herfa-Neurode repository.

Figure 2.4: Underground view of the Herfa-Neurode repository.

Figure 2.5: Underground view of the Herfa-Neurode repository.

layers, at a depth of about 800 m. Due to the clay layer, the salt deposit is isolated against the covering aquiferous layer and has therefore remained almost unchanged for the past 240 million years.

During the extraction of the potash deposits extending over an area of 1200 km^2, cavities were formed using the room and pillar mining method, which are now used for the disposal of hazardous waste materials. The underground waste disposal plant was admitted in accordance with waste law. The supervisory authority is the Mining Authority, Hessen. In addition to the Waste Act, mining regulations are also involved as far as the operation of the plant is concerned.

The capacity of the plant depends, practically, on the haulage capacity in the Herfa shaft, which has a payload of 7 tons. The annual capacity of the haulage plant is 200,000 tons. The underground cavities permitted by the mining authority for hazardous waste storage are sufficient for 20 more years. The 30% of the waste currently stored come from the local area of Hessen, 50% from other Federal Lands and 20% from foreign countries of Western Europe. The classification of the waste origin, and its percentage share of the total, is as follows:

- residues from the flue gas cleaning of incinerator plants: 30%;
- building rubble and earth excavation from demolition and renovation: 25%;
- metal-processing industry: 20%;
- residues from the chemical industry: 20%;
- electrical industry (transformers, capacitors): 5%.

The waste is put together into material groups. Within a material group, wastes which have similar substances are stored together.

2.4.3 Konrad iron mine

Iron ore mining started in the former Konrad mine (Fig. 2.6) in Lower Saxony in the sixties and was phased out for economical reasons in 1976 [22]. At the same year the Konrad site was selected for investigation as a possible repository because of the great depth of the ore horizon, the fact that the mine is extraordinarily dry and the complete isolation from shallow groundwater by clayish overlying rock.

Results of an extensive survey and evaluation programme led in 1982 to a positive statement regarding the site's suitability to host a radioactive waste repository. DBE has developed the repository technology, carried out the licensing procedure in cooperation with the government and will later transform the mine into a repository and operate it.

According to the license application, Konrad will be a repository for waste with negligible decay heat. Approximately 90% of the waste volume arising in Germany belongs to this category.

The Konrad repository will consist of 6 emplacement fields at different levels between 800 and 1300 m depth. A net disposal capacity of approximately 650,000 m^3 of waste packages will be available. Fig. 2.7 shows a scheme of planned mine operation.

Figure 2.6: The Konrad mine [24].

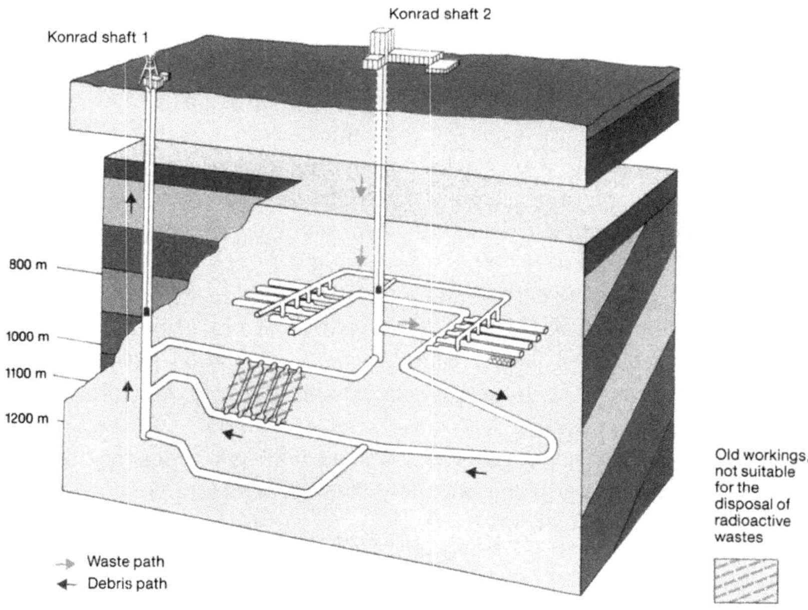

Figure 2.7: Scheme of planned mine operation [24].

2.4.4 Stripa iron mine

Mining of Stripa iron mine (Fig. 2.8) dates back to the 15th century. During long periods of time mining occurred only sporadically, with a complete standstill between 1634 and 1771 [25]. Mining ceased in 1976 with a total production of 18 million tons of crude ore – quartz banded hematite. The mining operation ceased because the whole orebody had been mined.

Between 1977 and 1980 a common Swedish-American project (SAC, Swedish American Cooperation) was carried through in Stripa. The project consisted of three main parts:

- heat experiments with simulated waste containers;
- evaluation of fissure hydrology;
- geophysical measurements.

Extensive information was obtained about mechanical reactions to heat in the control and ground water current in fissures in crystalline rock. The Swedish – American

Figure 2.8: Stripa mine [26].

Cooperation project attracted international interest and the international Stripa Project began in 1980. The research work was carried out as an independent project in the OECD Nuclear Energy Agency (NEA). Participating countries were: Finland, France, Japan, Canada, Great Britain, Spain, Switzerland, Sweden, and USA. The research was divided into the following areas:

- detection and mapping of fissure zones;
- groundwater conditions and nuclide migration;
- examination of bentonite clay for refilling and stopping up.

This part of the research went on up to the end of 1985. A third phase in the research began in 1986 and went on up to 1991. All previously mentioned countries except France and Spain participated in this part. The major aim of the third phase was research about:

- hydrogeology,
- chemical transportation,
- engineering barriers,
- geophysics.

2.4.5 Asse salt mine

Asse salt mine was used as a research laboratory for evaluation purposes (Fig. 2.9) of the salt disposal concept of Germany. The exploitation method used was room and pillar. The depth varies between 490 and 830 m. In 1965, the ownership of the Asse salt mine was transferred to GSF for the purposes of carrying out research

Figure 2.9: Storage of wastes in Asse mine for research purposes [27].

into the safe ultimate disposal of radioactive wastes. Since 1967, LLWs have been emplaced for experimental purposes until 1993, when experiments for the ultimate disposal of radioactive wastes at the Asse mine stopped.

Finally, it should be noted that there are also other underground mines that have been used as waste repositories, as shown in Table 2.4, but no detailed information about their operation is available. For example, chemical wastes have been stored in England since 1965 in an old mine at Walsall Wood colliery, at a depth of about 900 m. The mine is isolated environmentally by a geological graben with clay-filled faults on both sides and shale above.

References

[1] Kaliampakos, D. & Menegaki, M., Hazardous waste repositories in underground mines. A possible solution to an ever-pressing problem. *Proc. of the 1st Conf. On Sustainable Development & Management of the Subsurface*, Utrecht: Netherlands, 5–7 November 2003.

[2] Carmody, J. & Sterling, R., *Underground Space Design: A Guide to Subsurface Utilization and Design for People in Underground Spaces*, Van Nostrand Reinhold: New York, 1993.

[3] Kaliampakos, D., Mavropoulos, A. & Damigos D., Reducing risk of exposure from hazardous waste repositories, presented at the Environmental Health 2003 Conference, Catania, Italy, 2003.

[4] Kaliampakos, D., Mavropoulos, A. & Prousiotis, J., Abandoned mines as hazardous waste repositories in Europe. *Proc. of the 18th Int. Conf. On Solid Waste Technology and Management*, Philadelphia, PA, 23–26 March 2003.

[5] National Technical University of Athens (NTUA), *Survey of underground mines in Europe*. Low Risk Disposal Technology Research project (E.E. EVGI-CT-2000-00020), Deliverable D1.1, 2000.

[6] United States Geological Survey (USGS), *Minerals Information – Europe and Central Eurasia*, 2001, URL: http://minerals.usgs.gov/minerals/pubs/country/europe.html

[7] Knudsen, C., Nordic minerals review – Denmark, *Industrial Minerals*, No 374, pp. 52–55, November, 1998.

[8] Nurmi, A.P. & Peter, S.-W., Mining and Exploration in Finland, *Society for Geology Applied to Mineral Deposits, News*, No. 2, November 1996.

[9] Industrial Minerals, France – What next after MDPA has gone?. *Industrial Minerals*, No. 367, p. 54, April 1998.

[10] Mining and Metals, Tapping into Greece's mineral treasure chest, February 1998, URL http://www.ana.gr/hermes/1998/feb/mining.htm

[11] Sol, M.V., Peters, S.W.M. & Aiking, H., *Toxic Waste Storage Sites in EU Countries, A Preliminary Risk Inventory*, IVM Report number: R-99/04, February 1999.

[12] Geological and Mining Institute of Portugal, 2001.

[13] Mining Technology, 2000, URL: http://www.mining-technology.com

[14] Coal age, Bjorkdal gold mine is Europe's largest, *Coal Age*, **103(3)**, p. 38, March 1998.

[15] Beckius, K. & Thomaeus, M., Nordic review – a series of features highlighting the industrial minerals of Nordic countries. Sweden, *Industrial Minerals*, No. 374, pp. 52–82, November 1998.

[16] Industrial Minerals, Durham fluorspar mine closures imminent, *Industrial Minerals*, No. 373, p. 15, October 1998.

[17] Pearson, K., Potash producers. *Industrial Minerals*, No. 367, p. 57, April 1998.

[18] Industrial Minerals, McAlpine to extend Penrhyn slate quarry, *Industrial Minerals*, No. 365, p. 30, February 1998.

[19] Peila, D. & Pelizza, S., Civil reuses of underground mine openings: a summary of international experience. *Tunnelling and Underground Space Technology*, **10(2)**, pp. 179–191, 1995.

[20] Decamps, F. & Dujacquier, L., Overview of European practices and facilities for waste management and disposal. *Nuclear Engineering and Design*, Elsevier Science S.A., **176**, pp. 1–7, 1997.

[21] International Association for Nuclear Energy, 2001, http://www.uilondon.org

[22] DBE mbH, 2000, URL: http://www.dbe.de

[23] Kali und Salz, URL: http://www.kalisalz.basf.de

[24] Bfs, The Konrad Repository Project, From an Iron Mine to a Repository for Radioactive Wastes, Salzgitter, 1994.

[25] Stripa Mine Service AB, 1999, http://www.stripa.se

[26] Lawrence Berkeley National Laboratory, 1997, http://imglib.lbl.gov

[27] National Research Center for Environment and Health, 2000, http://www.gsf.de/Wir_ueber_uns/index_en.phtml

CHAPTER 3

Criteria for selecting repository mines

R. Pusch
GeoDevelopment AB, Lund, Sweden.

Abstract

This chapter discusses the most important criterion for selecting repository mines. The ideal mine to be converted to a safe waste repository is a rather small, modern, remotely and relatively deeply located, mine with intact power supply, pump systems and ventilation. The geological host medium is very important. Crystalline rock has excellent stability of the drifts and rooms even at large depths but it has a relatively high hydraulic conductivity. Salt contains no free water and offers very good isolation of the waste, but brine in local sediment lenses may cause difficulties in the preparation of the mine for waste application. Argillaceous rock has a very low hydraulic conductivity but poor stability and the vicinity of the drifts may be very conductive. The chapter gives examples of deep abandoned mines in granite, i.e. the Stripa mine in Sweden, formerly used for exploitation of iron ore, and of mined rooms in a salt dome as well as in limestone. They are taken as a basis of calculations of the mechanical stability, evolution of engineered barriers, and migration of released toxic elements to an imaginary well located close to the waste-filled rooms.

3.1 Introduction

The most important criterion for selecting repository mines is that the host rock should be low-permeable and mechanically stable, and that the mines should have suitable drifts and rooms for placement of waste packages. The ideal mine to be converted to a safe waste repository is a rather small, modern, remotely and relatively deeply located mine with intact power supply, pump systems and ventilation.

The geological host medium is very important. Crystalline rock has excellent stability of the drifts and rooms even at large depths but it has a relatively high hydraulic conductivity. The creep potential is very low and self-sealing and hence unimportant. Salt contains no free water and offers very good isolation of the waste, but brine in local sediment lenses may cause difficulties in the preparation of the mine for waste application. The creep potential is very high which means that the drifts and rooms converge and self-seal. Argillaceous rock has a very low hydraulic conductivity but poor stability and the vicinity of the drifts, i.e. the excavation-disturbed zone (EDZ), may be very conductive. It can undergo substantial creep and can self-seal depending on the diagenesis. Limestone rock has a low stability and is pervious. It has some creep and self-sealing potentials. Its most valuable property is the high pH of the porewater.

Mines in several other types of geological media, like metamorphous rock, shales and marble, can be considered for waste disposal but they are regarded here as representatives of the four types mentioned above. Thus, gneiss, shales and marble in principle behave like granite or argillaceous rock.

3.2 Rock structure

The structural constitution of the rock mass in which the mine is located determines the rate and distribution of the groundwater flow through the mine to the surroundings and hence the transport to the biosphere of toxic elements that can be released from the waste. For comparison and forming a basis for flow calculations it is necessary to work out relevant generalized rock structure models.

The transmissivity of the host rock determines its isolating capacity and is therefore the most important factor for the long-term function of the mine. It is controlled by the rock structure, which is hence a primary factor for the performance. It is also a determinant of the mechanical stability of the mine. The scheme in Table 3.1 is applicable to most rock types except salt.

3.2.1 Crystalline rock

Many granites can be represented by a generalized orthogonal-type rock structure model like the one shown in Fig. 3.1, which includes practically important discontinuities of different orders. Metamorphic rock like gneiss commonly has a more wavy macroscopic nature and more anisotropic structural organization but it can still be represented by the same model except that the spacing of sixth- to fourth-order discontinuities is usually much smaller in one direction than in the two other. We will consider a deeply located underground research laboratory in Sweden, extending from a former iron ore mine in Sweden, as the reference case (Stripa mine) [2].

3.2.2 Argillaceous rock

Mines in argillaceous rock are common in many countries, the exploitable ore often being impregnations of lead, copper and a number of other metals (Fig. 3.2).

Table 3.1: Categorization scheme for rock structure with typical geometrical, hydraulic and strength data [1, 2]. (The crystal matrix has typically $K < 10^{-13}$ m/s.)

Discontinuity	General properties	Length	Width (m)	Hydraulic conductivity (K) (m/s)
Low-order discontinuities (conductivity of respective discontinuity)				
First order	Very large faults, water-bearing, gouge, very low strength	>Kilometres	Hundreds	10^{-7}–10^{-5}
Second order	Major fracture zones, water-bearing, gouge, low strength	Kilometres	Tens to hundreds	10^{-8}–10^{-6}
Third order	Minor fracture zones, water-bearing, some gouge, relatively strong	Hundreds of metres	Metres to tens of metres	10^{-10}–10^{-7}
High-order discontinuities (conductivity refers to rock with no discontinuities of lower order)				
Fourth order	Discrete major fractures, water-bearing, little or no gouge, strong	Tens of metres	–	10^{-11}–10^{-9}
Fifth order	Little water, no gouge, high strength	Metres	–	10^{-12}–10^{-10}
Sixth order	Decimetres, no water, no gouge, very high strength	Decimetres	–	10^{-13}–10^{-11}
Seventh order	Centimetres and smaller (fissures, voids)	Centimetres and smaller	–	$<10^{-13}$

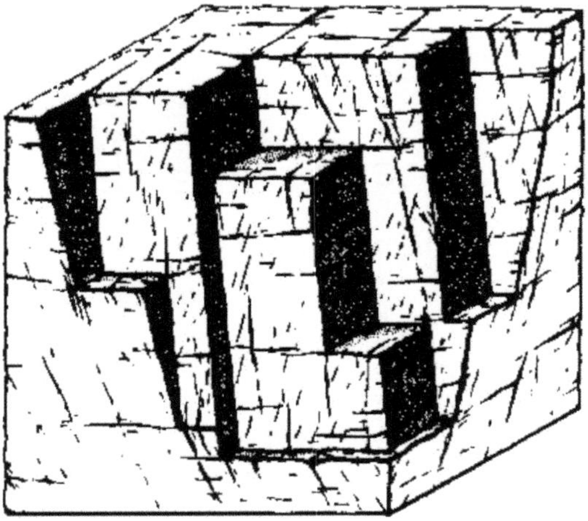

Figure 3.1: Typical appearance of crystalline rock. The breaks represent third- and fourth-order discontinuities while the finer weaknesses represent fifth- and higher-order discontinuities.

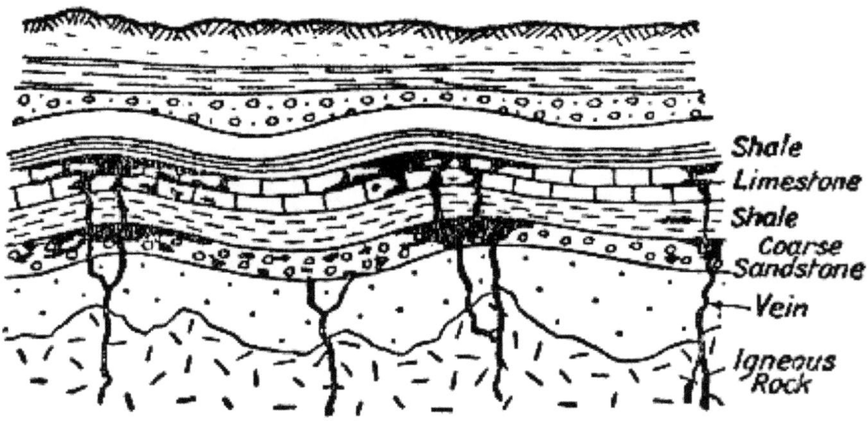

Figure 3.2: Formation of ore (black) in sedimentary rock due to impregnation by solutions.

The figure is typical in the sense that the mode of formation, i.e. deposition and consolidation of sediments, has often resulted in very strong variation in composition and geotechnical properties.

Argillaceous rock can be considered as a less brittle version of metamorphic rock and can also be represented by Fig. 3.1 although with even smaller spacing of the discontinuities in one plane (direction), and hence significant anisotropy.

Comprehensive experience from exploration of such rock in France and Switzerland has shown that low-order discontinuities are less conductive than in crystalline rock because tectonically induced shearing has led to disintegration and significant sealing. The rock matrix, i.e. the rock between the low-order discontinuities, has a hydraulic conductivity and a transmissivity as low as those of crystalline rock or somewhat lower.

Drifts and tunnel systems in northern Switzerland serve as typical representatives of rooms for disposal of hazardous waste in sedimentary rock. Such rooms are presently used as underground research laboratories for development of techniques for disposal of highly radioactive waste (Mont Terri), [3].

3.2.3 Salt rock

Rock in the form of domal salt – usually of sodium or potassium type – or bedded salt is very suitable for waste disposal except if it contains brine pockets. It is already being utilized in many countries like Germany and France because completely homogeneous salt is perfectly tight with respect to water and gas. A necessary prerequisite is to construct long-lasting seals in the form of plugs in the shafts leading down to the repository level since water inflow from shallow soil and rock can cause very difficult problems. Salt mining has been made at all levels but only mines located several hundred metres below the ground surface should be considered.

Figure 3.3 is a schematic drawing of a salt dome, surrounded by other rock that caused salt in an underlying salt bed to be squeezed up to form the dome. The process forced up material from the surroundings yielding internal flow structures and irregularly spaced and oriented lenses of clay/silt/sand, which represent inclusions of brine. All these discontinuities affect the stability of rooms in the salt rock and can contain brine. The figure also shows a cross section of the Asse salt mine that has been used as an underground research laboratory by the Gesellschaft fuer Anlagen- und Reaktorsicherheit GmbH (GRS) in the last decades [3]. It may well be used for disposal of hazardous chemical waste. The spacing of major discontinuities varies with the size of the salt domes and for the Asse case it is assumed to be 50–100 m within 200 m distance from the boundary of the dome and 100–200 m in the interior. They do not intersect, and hence the only connection with the biosphere is where the rooms are intersected by lenses that extend all the way up the ground surface. The lognormal persistence of the lenses is assumed to vary between 100 and 1000 m, the shorter ones being termed here third-order discontinuities and the longer ones second-order discontinuities.

Despite the excellent isolating potential of salt caused by the absence of free water, there are two not solved problems yet: (1) gas production in the waste saturated with brine can cause extremely high pressures that may lead to upward penetration of gas and expulsion of heavily contaminated brines; (2) the very significant creep properties of salt rock will make retrieval of waste packages impossible after some 50–100 years since heavy objects sink in an unforeseeable

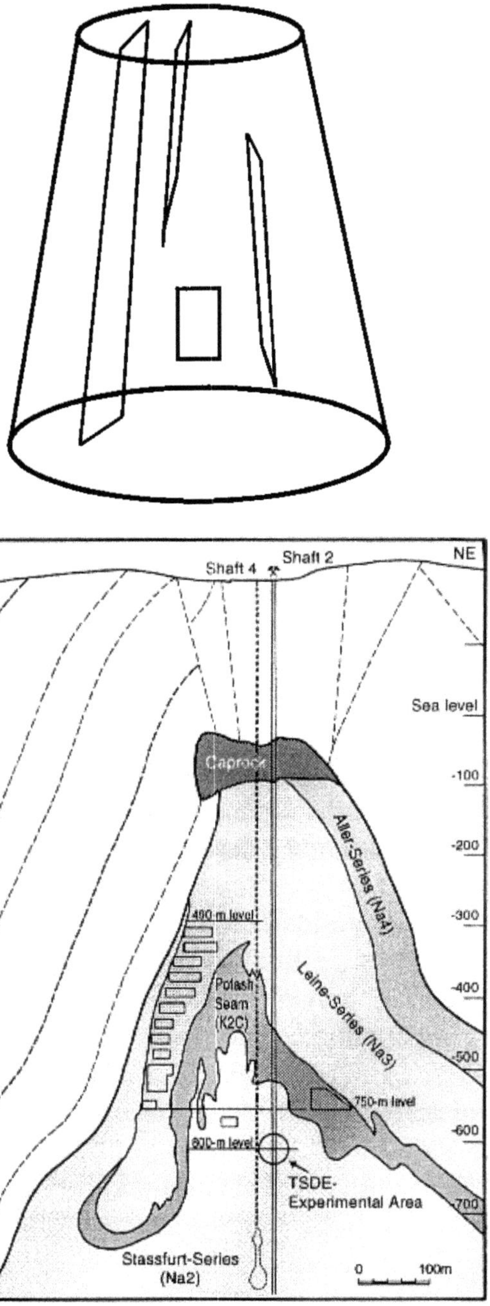

Figure 3.3: Structural nature of a salt domes. Top: Generalized rock structural model. Bottom: The Asse salt mine [3].

way in the salt mass. For these reasons disposal in salt cannot be proposed as a first choice method.

3.2.4 Other rock types

In certain countries mining has been extensive in very porous rock like limestone and abandoned mines in such geological media may be the only alternative. Mines located in regularly bedded limestone as well as heterogeneous limestone like ancient coral reefs, may have to be considered as options for disposal of hazardous waste. The hydraulic conductivity and transmissivity may be very high and the mechanical stability very low, which is of course not suitable. However, the high pH of the groundwater provides good chemical conditions for minimizing dissolution of various types of waste. The permeable nature of many limestone regions requires construction of very effective engineered barriers. We will use a bauxite mine in limestone environment in Greece as a reference in later chapters dealing with stability and waste isolation efficiency.

3.3 Requirements for the use of mines as repositories

3.3.1 Function of the host rock

A major criterion for use of abandoned mines as waste repositories is that elements released from the waste must not contaminate groundwater in the mine area more than what is accepted by regulatory authorities. The geological medium must provide mechanical protection of the 'chemical apparatus', i.e. the backfilled rooms with waste embedded in and surrounded by engineered barrier systems (EBS), and yield slow release of toxic elements to the biosphere. The processes that can threaten and degrade the geological medium hosting the repository mine are tectonic events, glaciation involving deep abrasion and erosion, and loss of sealing ability of the engineered barriers.

The capacity of mine repositories to isolate waste is determined by:

1. The physical stability of the rock and the physical and chemical performance as well as the stability of the EBS.
2. The rock structure and related hydraulic conductivity and the conductivity of the EBS are key factors for the performance of mine repositories.
3. The impact of groundwater chemistry on the longevity of the EBS, and the chemical nature of dissolved hazardous waste elements released from the EBS are key factors.

3.3.2 Conversion of mines to repositories

The major issues are the location of the mine with respect to the risk of contaminating drinking water in the closest wells, the status of the mine with respect to the mechanical stability and the cost for converting it into a repository. The most important parameters in addition to the issue of groundwater contamination are:

- size,
- remaining exploitable ore,
- rock structure, hydrology, and stability,
- transport to and in the mine,
- technical facilities,
- stabilization,
- cost.

3.3.3 Size

Mines that can be considered for disposal of toxic chemical waste are those with relatively large underground space suitable for storing the waste, 5000 m^3 being a practical minimum. Drifts for disposal should have a geometry that is suitable for rational application of waste packages. Horse-shoe or rectangular cross section shape of drifts and rooms with 5–30 m height and width and a length of 50–200 m are preferable. Big rooms may require significant stabilization and sealing of the rock.

3.3.4 Remaining exploitable ore

The reasons for abandoning a mine is commonly that practically all the ore has been mined out or that continued extraction of ore gives too little profit. However, in the second case, future methods for extracting and refining valuable minerals may make such mines interesting again. Thus, careful analysis has to be made of what the possibilities are to continue mining operations. Metal ore may be of quite different economic value today and in a few tens of years from now, of which uranium, titanium, gold and platinum are historical examples. It is much less significant for iron ore mines, which therefore represent attractive alternatives. Other mines of considerable interest are those in which clay minerals have been mined. An attractive principle in selecting mines for disposal of chemical waste is to use mines in which the exploited ore contained the same types of hazardous elements as the waste. Thus, mines in sulphide ore districts where mercury, arsenic, and lead have impregnated the rock and contaminated the groundwater are suitable for disposal of waste containing these elements.

One can also include abandoned, deeply located railway and road tunnels in the group of fully exploited mines. Examples are certain road tunnels in the Alps region located kilometres below the ground surface.

3.3.5 Rock structure, hydrology, and stability

3.3.5.1 General

The suitability of an ore mine to be used for waste disposal depends primarily on the risk of contamination of the groundwater, especially water for drinking purposes and irrigation. Most rocks are permeable because of the presence of natural fractures and systems of fracture zones, which determine the transport of contaminants and make certain host rocks less suitable. An important fact is that the drifts and mined-out rooms are surrounded by an excavation-disturbed zone (EDZ) with reduced mechanical stability and increased hydraulic conductivity. Where the EDZ interacts with natural strongly water-bearing fracture zones there are conditions for poor stability and quick and extensive transport of contaminants to the biosphere. This is the reason why all waste disposed in any underground repository must be effectively isolated by engineered barrier systems. The transmissivity of a rock mass is much higher when the frequency of water-bearing discontinuities is high than if the spacing of such features is low and their interconnectivity poor. Even the first mentioned type of rock can be used successfully for safe waste disposal but more effort and money have to be spent on the engineered barriers than in the latter case. Hence, stable and low-permeable rock is preferable.

It is desirable to select mines with neutral or slightly alkaline ground-water because it minimizes the solubility of heavy metals and enhances the longevity of cement- and clay-based engineered barriers provided that pH is not too high.

3.3.5.2 Rock structure modelling

Rock structure is a key issue since reliable calculation of the transport of chemical elements released from the waste requires that a representative rock structure model can be defined and used. The matter has been considered in several practical hydropower projects and comprehensive research programmes related to the disposal of radioactive waste. In this book we will show how rock structure models can be used for both rock mechanical and hydrological calculations. The basis is the rock categorization scheme in Table 3.2, which is a simplified version of Table 3.1. Figure 3.4 shows a generalized structural model.

Table 3.2: Generalized scheme of rock discontinuities [1]. First to third are fracture zones, fourth to seventh are discrete fractures.

Order	Length	Hydraulic conductivity	Gouge content	Strength
First	>Kilometres	VH	VH	VVL
Second	Kilometres	H	H	VL
Third	Hundreds of metres	M to H	M	M
Fourth	Tens of metres	M to H	L	H
Fifth	Metres	L	VL	H
Sixth	Decimetres	VL	VL	VH
Seventh	<Decimetres	VVL	VVL	VVH

VH, very high; H, high; VL, very low; L, low; M, medium; VVL, insignificant; VVH, very very high.

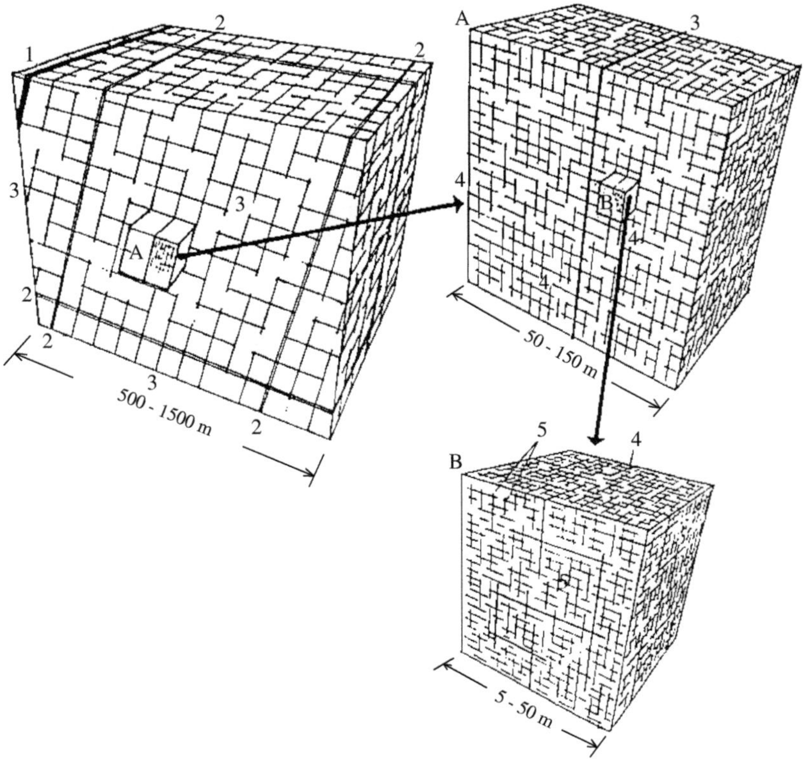

Figure 3.4: Generalized rock structure model for hydraulic and mechanical calculations. The patterns are self-repeating with respect to geometry but not to the physical properties and hence do not represent true fractals.

The stability of the host rock is of importance since tectonic events will cause shearing of discontinuities of fourth and lower orders, affecting their hydraulic conductivity and the stability of drifts and rooms. For the rock matrix with only fifth- and higher-order discontinuities, Griffith's or other failure criteria apply and fracture mechanics must be used for calculation of fracture strain and propagation. Furthermore, creep will take place that can cause critical strain and failure of the near-field rock field of drifts and rooms.

3.3.6 Transport to and in the mine

Transport of waste material to the mine should be simple and safe. Railway transport is preferable and most mines have such facilities. The issue is not critical but if a choice between otherwise equally suitable mines has to be made, access to good and safe ways of waste transport is an advantage. Transport of toxic chemical waste for storage in deep mines should be on land while ship transport on rivers or near the coast should be avoided or – in sensitive areas - not allowed.

Almost all mines have railway connections, which is naturally valuable for transport of waste for disposal and for transport of material in the reconstruction phase. Mines with internal transport facilities like decauville railways and elevators are preferable especially if they are intact or need only little repair and maintenance.

Preparation of the waste, like solidification of liquid waste, casting of solid waste in cement, and compaction of clay packages containing dispersed solid waste, can be made in buildings on the ground surface in the mine area but can also be performed down in the mine. The latter way is advantageous since the transport distance will be smaller and the risk of damaging containers and packages hence lower, but the space required may disqualify small mines.

3.3.7 Facilities and installations

Modern mines have safe electric power supply and good ventilation and pump systems for drainage. If these facilities remain at the time of converting the mine into a repository it implies substantial cost saving.

3.3.8 Stabilization

Old mines represent dangerous conditions with unstable rock and failed anchorings and comprehensive stabilization may be required. This again speaks in favour of utilizing modern mines or even better mines that are still in operation.

A major problem is represented by the risk of accumulation of explosive gas. This is the case for coal mines, which may be considered for waste disposal if the coal seams are part of cyclothemes with smectite clay as one component. Very significant safety measures have to be taken and such mines are not suitable especially since permeable sandstones are often associated with the clay.

An important issue is the long-term performance of mine repositories with respect to significant internal strain due to rock stresses that prevail or imposed by tectonics and glaciation. The effect can be difficult to predict and may imply that certain rocks like argillites should not be used. The usually good mechanical stability of crystalline rock makes it a good candidate despite the presence of water-bearing fracture zones. Salt rock is also suitable since internal strain, which will be very significant, ultimately leads to convergence and self-sealing because of the extreme creep properties.

3.3.9 Cost

It is of utmost importance that the cost for safe disposal of toxic chemical waste in abandoned mines be kept as low as possible. Thus, while disposal of radioactive waste is part of the nuclear fuel cycle and budgeted from start, the cost for handling and disposal of chemical waste has to be carried by the society or the waste-producing companies, i.e. in practice by the tax payers. The most expensive activities and facilities in preparing abandoned mines for waste storage and in applying waste are (1) backfilling of ramps and shafts and isolation of the waste, (2) stabilization and sealing of the mine, (3) pump systems, drainage and ventilation, and (4) elevators.

3.4 Reference mines

3.4.1 General

We will define reference cases, representing crystalline, argillaceous, salt and limestone rock for illustrating what sort of data and calculations that are required in performing some of the design work required for converting abandoned mines into repositories for hazardous chemical waste. Calculations will be made mainly for crystalline rock since it represents a common case and special difficulties. A further reason is that argillaceous rock and limestone of sedimentary origin often show similar structural features.

3.4.2 Crystalline rock

3.4.2.1 The Stripa Mine

The reference case is the abandoned Stripa iron ore mine that reaches down to 410 m depth in a granite dome. Iron ore has been exploited in this mine from the year 1400 to about 1980. It was then used by the Swedish Nuclear Fuel and Waste Management Co (SKB) and Lawrence Berkeley Laboratories, California, as an underground laboratory for testing techniques for isolating nuclear waste, later in the form of the international Stripa Project [2]. The mine is presently owned by a waste handling company (Ragn-Sell AB) and considered as a possible repository for Hg disposal.

As outlined earlier in the book the rock structure and rock stress conditions are of major importance and they will be described here for this mine. Rock mechanical calculations for assessing the stability of typical drifts and rooms for waste disposal and for getting a basis for designing stabilizing constructions, and hydrological calculations for determining the risk of contamination of a hypothetic well for drinking water, will be described in subsequent chapters.

3.4.2.2 Regional rock structure

The granitic region where the Stripa mine is located is characterized by first- and second-order discontinuities oriented and spaced as in Fig. 3.5. On this large scale the major low-order discontinuities make up two steep NW-SE and NE-SW oriented sets. Close examination of finer weaknesses has shown that those of third and fourth orders also have these orientations and that there is one more set of breaks on all scales that is more or less subhorizontal.

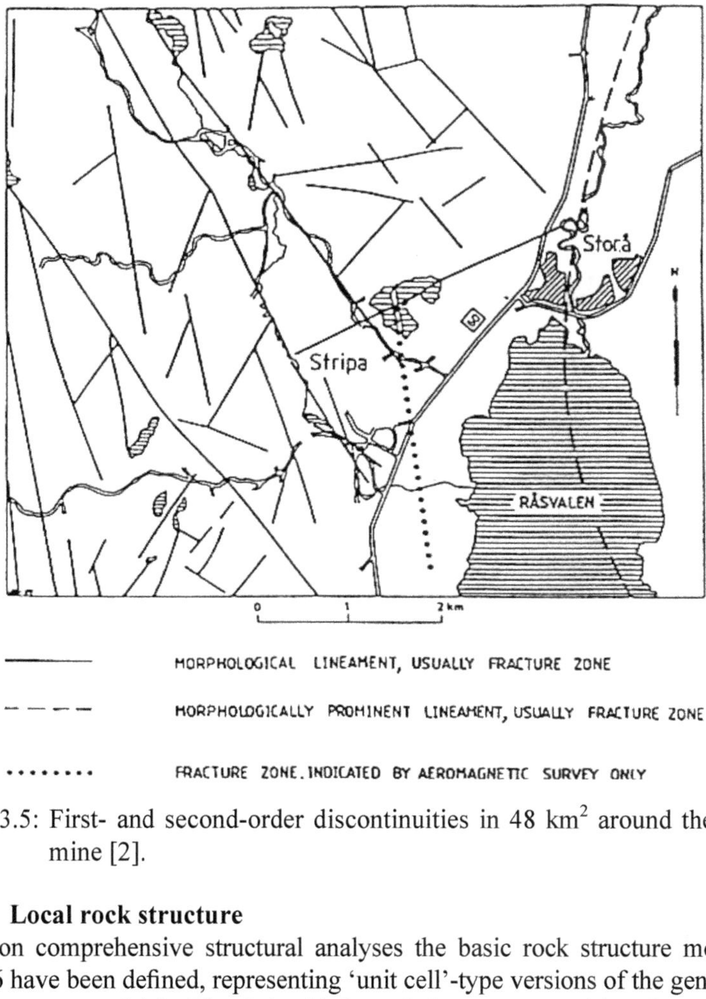

Figure 3.5: First- and second-order discontinuities in 48 km^2 around the Stripa mine [2].

3.4.2.3 Local rock structure

Based on comprehensive structural analyses the basic rock structure models in Fig. 3.6 have been defined, representing 'unit cell'-type versions of the generalized rock structure model in Fig. 3.4 with boundaries represented by discontinuities of second and third orders, respectively. For the various hydrological and mechanical modelling attempts the physical properties at the boundaries are given in Tables 3.3 and 3.4.

The rock structure model in Figs 3.4 and 3.6, with all discontinuities grouped orthogonally, applies in principle to the Stripa area. The following generalized geometrical data of third- and fourth-order discontinuities can be applied for rock mechanical and hydrological modelling:

Third-order discontinuities Steeply oriented third-order discontinuities with 75 m spacing conform to the second-order zones. The actual spacing is about 50–100 m and typical of many granite bodies.

Fourth-order discontinuities Most of the fourth-order discontinuities conform to the third-order discontinuity sets. Their spacing is in the range of 2–4 m and they persist for several tens of metres.

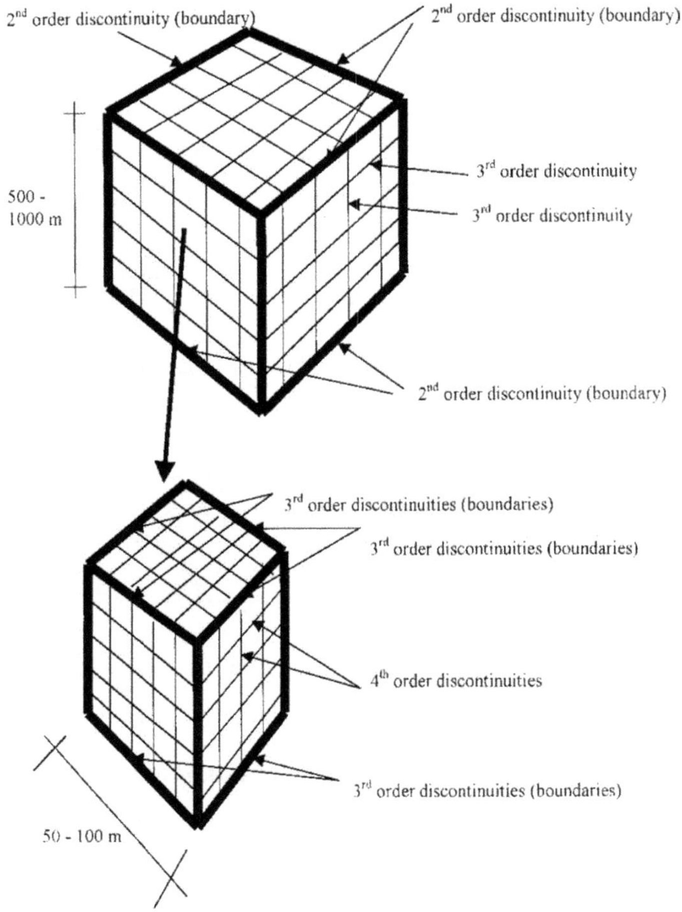

Figure 3.6: Rock structure model for the Stripa host rock. Spacing of second-order discontinuities (fracture zones) 500–1000 m. Spacing of third-order discontinuities 50–100 m. Spacing of fourth-order discontinuities 5–10 m.

3.4.2.4 Rooms

The Stripa mine has a number of drifts and tunnels and rooms of various size. Several of them are suitable for waste disposal and one room and one drift, have been selected for the modelling work that is described in subsequent chapters. With some generalization they can be defined as a drift with 25 m^2 horseshoe-shaped cross section, and a room with 50 m width, 50 m height, and 100 m length (Fig. 3.7). The centre of both are assumed to be located at z = 360 m. Thus, in the cubical rock element with 600 m edge length that contains the mine rooms (x = 600 to 1200 m, W-E direction, y = 100 to 700 m, S-N direction, and z = 0 to 600 m, from ground surface downwards) there are three second-order discontinuities.

Table 3.3: Assumed physical properties of discontinuities and rock matrix for crystalline rock [1, 2].

Rock	Hydraulic conductivity (m/s)	Transmissivity (m^2/s)	Mohr/Coulomb friction angle, ϕ°	Mohr/Coulomb cohesion, c (MPa)
First-order discontinuities	10^{-7}–10^{-5}	10^{-5}–10^{-2}	15–20	0
Second-order discontinuities	10^{-8}–10^{-6}	10^{-7}–10^{-4}	20–25	0
Third-order discontinuities	10^{-9}–10^{-7}	10^{-9}–10^{-6}	20–30	0
R4	10^{-11}–10^{-9}	–	20–35	0.1–1
R5	10^{-12}–10^{-10}	–	35–50	1–10
R6	10^{-13}–10^{-11}	–	45–60	10–50

R4, R5 and R6 represent rocks with discontinuities finer than fourth, fifth and sixth orders, respectively.

Table 3.4: Mohr/Coulomb parameter data as functions of the size of the crystalline rock volume [1].

Rock volume (m^3)	Cohesion (MPa)	Peak friction angle (°)	Discontinuities in the rock volume
<0.001	10–50	45–60	<Sixth order
0.001–0.1	1–10	40–50	Sixth order
0.1–10	1–5	35–45	Fifth, sixth order
10–100	0.1–1	25–35	Fourth, fifth, sixth order
100–10000	0.01–0.1	20–30	Third, fourth, fifth, sixth order
>10000	<0.1	<20	All

The actual shape of the tunnel means that the roof is curved with 5 m radius, while the walls are vertical and 5 m apart. The distance between the flat floor and the crown is 5 m.

Excavation disturbance extends to 1 m distance from the periphery of the drift and to 3 m distance from the big room.

3.4.2.5 Rock stress conditions

In general the vertical rock pressure represents the minor principal stress and equals the overburden pressure. Typical primary stress conditions according to measurements (mainly door-stopper and hydraulic fracturing) are illustrated in Table 3.5.

Measurements in Stripa at 360 m depth have given σ_H = 15–30 MPa, σ_h = 5–15 MPa and σ_v = 10 MPa. The average values σ_H = 23 MPa, σ_h = 12 MPa and σ_v = 10 MPa are used in the rock mechanical modelling.

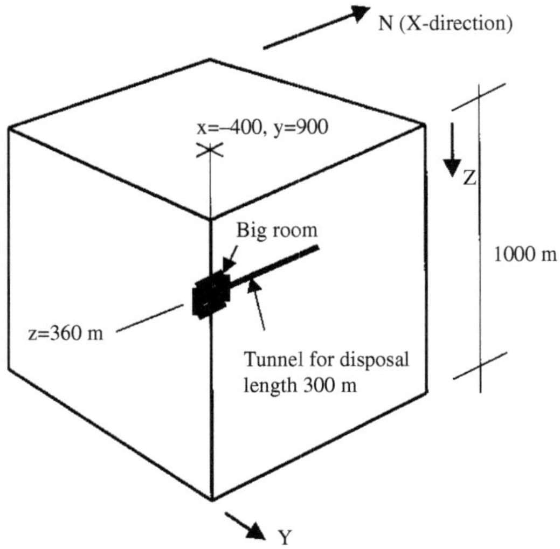

Figure 3.7: The Stripa room and tunnel.

Table 3.5: Magnitude of rock stresses at Stripa [1].

Depth, z (m)	Vertical stress (MPa)	Maximum horizontal stress (MPa)	Ratio of maximum and minimum horizontal stress
0–900	$10 + 0.027z$	$10 + 0.037z$	2–4
900–2200	$33 + 0.01z$	$35 + 0.01z$	1–2

3.4.2.6 Rock stability issues

The following processes are of importance:

- block fall from roofs and walls,
- overstressing of the periphery of the rooms,
- blasting (excavation disturbance).

Critical constellations of fourth-order can cause unstable rock wedges that may drop down, and rock rich in fifth- and higher-order discontinuities may cause comprehensive rock fall. This requires analysis of rock structure models and securing of potentially unstable blocks. Too high hoop stresses may cause breakage and rock spalling. Such breakage and blasting-induced disturbance causes the 'excavation-disturbed zone'. This zone has a high hydraulic conductivity and serves as an effective flow path.

The big caverns of the Stripa mine are known to have a poor stability because the EDZ extends to about 5 m from the walls and at least 2 m from the roof and comprehensive rock fall has taken place due to critical fracture constellations and hoop stresses. The statistical risk of fall of big blocks can be determined by using the rock structure models.

A basis for investigating whether critically high hoop stresses will cause breakage and what the associated extension of the EDZ will be is determined by the compressive strength of the rock. It can be taken according to Table 3.4. Other major rock material data are taken to be: Young's modulus, $E = 50$ GPa, Poisson's ratio = 0.20, Creep law: $\varepsilon = \sigma^n/E + (\sigma^n/\eta)t^\alpha$, with $n = 3$, $\eta = 10^{18}$ Pa s, and $\alpha = 0.3$.

3.4.2.7 Hydrology in the far-field and near-field

In general, the far-field hydrology is determined by second- and third-order discontinuities and the continuous EDZ that short-circuits the discontinuities. The near-field hydrology is determined by the fourth-order discontinuities and the EDZ that forms a continuum and short-circuits the discontinuities.

In the Stripa mine the undisturbed granite mass that is confined by third- and lower-order discontinuities has an average hydraulic conductivity of 10^{-11} m/s as evaluated from careful inflow tests in tunnels [1]. For second- and third-order discontinuities Table 3.3 applies.

It is estimated that the EDZ around the periphery of the big caverns extends to about one-fifth of the width of drifts and rooms. The hydraulic conductivity of the EDZ is in the range of 10^{-9}–10^{-6} m/s; the average value can be taken as 10^{-7} m/s for the big room. The blasting-induced EDZ around the drifts and tunnels extends to about 0.5 m from the walls and roof and to 1.5 m from the floor. Its hydraulic conductivity is 10^{-9}–10^{-8} m/s.

3.4.3 Salt and argillaceous rock

The present project includes consideration of salt and clayey sedimentary rock for disposal of hazardous waste but focus is on crystalline rock in which the reference disposal site, the Stripa mine, is located. No selection of corresponding typical mines in salt and argillaceous rock has been made but for illustrating the different conditions for adapting such mines to repositories some general material data are summarized in Table 3.6.

Table 3.6: Typical physical–chemical properties of the host media and the engineered barrier system. The top value in each cell refers to the near field and the second to the far field.

Property	Granite	Salt (halite)	Clay shale	Limestone
Bulk density (kg/m^3)	2650–2700	2200	2500–2600	2200–2700
Total porosity	0–2%	<<0.01	1–5%	5–20%
Surface area (m^2/ton)	1×10^2	0	1×10^4	10^3
Tortuosity	10^{-6}–10^{-4}	–	10^{-1}	10^{-1}
Organic carbon fraction (f_{oc})	0	0	0.1–5%	1–10%

References

[1] Pusch, R., *Rock Mechanics on a Geological Base*, Developments in Geotechnical Engineering, Vol. 77, Elsevier: New York, 1995.
[2] Pusch, R., *Waste Disposal in Rock*, Developments in Geotechnical Engineering, Vol. 76, Elsevier: New York, 1994.
[3] Svemar, Ch. & Pusch, R., CROP – Cluster Repository Project: A basis for evaluating and developing radioactive concepts of final repositories for high-level radioactive waste. Final Report EC-contract FIR1-CT-20007-200023, Brussels, 2005.

CHAPTER 4

Engineered barriers

R. Pusch
GeoDevelopment AB, Lund, Sweden.

Abstract

This chapter presents the results of the survey of materials for constructing engineered barriers in mine repositories. Engineered barriers for the isolation of solid or solidified waste for disposal in abandoned mines must be chemically compatible with both the waste and the rock. While metal containers can be a part of the multi-barrier system, they turn out to be less suitable both for cost reasons and for the reason that most types generate gas that causes problems from a safety point of view (hydrogen gas). This chapter describes the selection procedure, focusing on cement and clay, for use in mine repositories in crystalline rock, argillaceous rock, and limestone. For backfilling of rooms and drifts in salt, crushed salt is a major barrier candidate. Among clay materials those rich in smectites have the best isolating properties but they also represent the highest costs. A clay material that appears to represent an optimum with respect to cost and good isolating properties is the German Friedland Ton, which was used in the study for modelling the performance of clay barriers. The fact that initially dry clay material must become almost completely fluid saturated before diffusive migration can start means that the time for hazardous species to appear in remote wells is further delayed. Thus, depending on the thickness of engineered barriers in the form of compacted smectitic clay it can take centuries or even thousands of years before diffusive ion transport will become significant. All transport processes in clays are controlled by their density, which determines the microstructural constitution, and by the type of cations and anions. The most important physical properties of smectite clays that make them suitable for isolation of hazardous waste are the ability to undergo strain without fracturing, the very low hydraulic conductivity, the gas penetrability, the cation sorption capacity, and the ability to expand and establish tight contact with the waste and the rock. Cement is used for underground construction work in mine repositories where it will be in contact with the clay barriers. They are not compatible with respect to chemical interaction and earlier investigations have shown that low-pH cement performs better than Portland cement. The present study comprised laboratory testing of Friedland Ton in cells with cement porewater contained in centrally placed tube filters. This water, which represents the water phase of water-saturated low-pH Spanish and Swedish cements, ran for about one year and showed that the chemical interaction with the clay did not lead to any significant destruction of either.

4.1 Types and characteristics of engineered barriers

The main focus has been on clay and cement for use in mine repositories in crystalline rock, argillaceous rock, salt, and limestone. Concrete plugs have to be constructed in strategic positions for confining the engineered barriers and for redirecting groundwater flow.

4.1.1 Clay

4.1.1.1 Fundamental behaviour of clay/water systems

The electrical charge and colloidal size of clay particles make them hydrate and interact such that their hydraulic conductivity and stress/strain properties of clays are quite different from friction soils. This is particularly important for smectites because their net charge is higher than that of any other clay minerals. Their fine tortuous pore systems offer considerable resistance to water and gas flow, especially at high densities and they absorb and bind both water and cations as well as small amounts of anions. The soil-water potentials are of fundamental importance to engineered clay barriers [1].

The energy status of the clay-water system governs the development of 'matric' and 'osmotic' forces that are responsible for hydration and dehydration. In general one can explain clay water suction that is measured by tensiometers etc as the net effect of the thermodynamic terms:

- Ψ_m = matric potential;
- Ψ_s = osmotic, i.e. solution potential (referring to the interaction between solute and water molecules); $\Psi_s = nRTc$, where n = number of molecules per mole of salt, R = universal gas constant, T = absolute temperature and c = concentration of solutes;
- Ψ_g = gravitational potential;
- Ψ_a = pneumatic air pressure;
- Ψ_p = external pressure transmitted to the particles through the fluid phase.

It is generally accepted that no more than three interlamellar hydrate layers can be established in any smectite clay species. While the stacking of lamellae at this and smaller distances, which correspond to high bulk densities, can be approximately taken as a plane-parallel arrangement of particles it is quite different at lower swelling pressures and the energy state of the porewater is consequently not the same. For lower bulk densities a large part of the porewater volume is 'free' and located in voids between particle stacks and aggregates, which is the term for closely located cohering stacks of lamellae.

The type of absorbed cations is important since bivalent and polyvalent cations cause growth in stack thickness and size, which means that for any density the voids between the stacks of lamellae are bigger when calcium and other multivalent cations are adsorbed than for sodium, which is the natural adsorbed cation in Friedland Ton.

The amount of interlamellar water exceeds that of free porewater at higher densities and since the first mentioned fraction is immobile at normal hydraulic gradients the hydraulic conductivity of dense smectitic clay is very low irrespective of the type of adsorbed cation. A high density of waste-embedding clay is hence asked for.

Interaction of electrical double-layers at the external surfaces of stacks of lamellae is an osmotic phenomenon. Figure 4.1 describes the charge distribution in the narrow space between two adjacent parallel smectite particles with a spacing that allows development of complete electrical double-layers. This matter is basic to the understanding of ion diffusion, which is discussed later in this chapter.

The implication in Fig. 4.1 is that the space between clay aggregates is smaller at higher densities and that the equally charged units repel each other and hence contribute to the swelling pressure. The net effect of porewater salinity and bulk density on the microstructure is illustrated by Fig. 4.2, which also shows the relation to the swelling pressure. For calcium and other multivalent adsorbed cations the much fewer contacts between stacks, i.e. the lower frequency of interacting external double-layers, means that the influence of the salt content on the swelling pressure is appreciably smaller than for the sodium case at low and

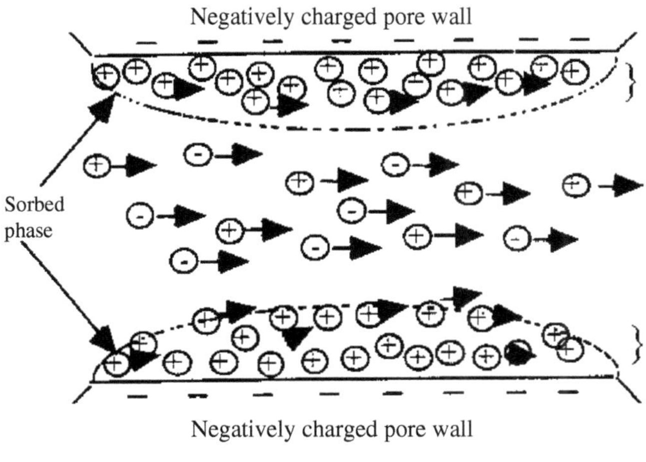

Figure 4.1: Schematic picture of interacting electrical double-layers. Brackets denote the sorbed phase in which surface diffusion takes place.

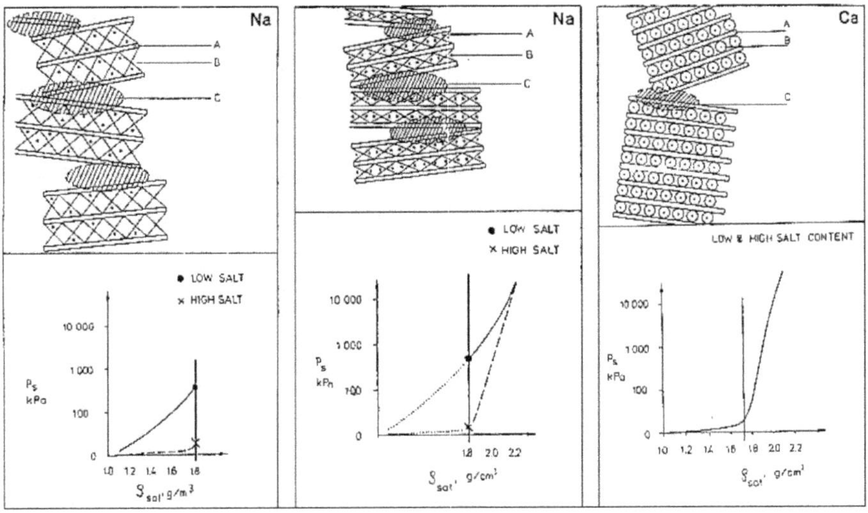

Figure 4.2: Schematic pictures of stack assemblages and influence of density at water saturation, expressed in g/cm^3 (1 g/cm^3 = 1000 kg/m^3) and salinity for sodium and calcium montmorillonite clay. (A) lamella, (B) interlamellar space, and (C) stack contact region with interacting electrical double-layers [1].

intermediate bulk densities. Hence, there is a difference in swelling behaviour explained by microstructural differences.

'Intracrystalline' swelling takes place through the matric potential Ψ_m and the osmotic potential Ψ_s. The presence of air/water interfaces is not required for water uptake caused by the matric potential. Hence, fully water-saturated smectite clay will absorb water and expand until the hydration capacity for any given density is fully used up. On contacting smectite clay with water, like in the case of clay blocks filling a drift in a mine, water will be sucked up from the rock by the negative porewater pressure that is set up and the wetting rate will be of diffusion-type unless the water pressure is significant (>1 MPa) in which case flow may also be significant. The negative pressure is of the same order of magnitude as the swelling pressure, which depends on the density, up to several MPa.

4.1.1.2 Clay materials for waste isolation

Table 4.1 shows examples of commercial smectite clay materials for clay embedding of solid waste.

The most important physical properties of smectite clays that make them suitable for isolation of hazardous waste are:

- ductility – ability to undergo strain without fracturing;
- hydraulic conductivity – very limited permeation by water;

Table 4.1: Examples of commercial smectite clays and the major sorbed cation in the natural clay [2]. They can all be transformed to sodium form by soda treatment on an industrial scale.

Clay	Manufacturing company	Smectite content
Montm. Tixoton, Ca	Süd-Chemie (Germany)	90
Montm. RMN, Ca	Obrnice (Czech Republic)	90
Montm. IBECO, Ca	Silver & Baryte (Greece)	80
Montm. MX-80, Na	American Colloid (USA)	75
Montm. Moosburg, Ca	Süd-Chemie (Germany)	65
Montm. Kunigel, Na	Kunimine Ind., Japan	50
Saponite S, Ca	(Through Enresa, Spain)	70
Beidellite, Ca	(Through Enresa, Spain)	35
Mixed-layer, Friedland	Frieton (Germany)	45

- gas penetrability – gas penetration at low pressure excludes risk of highly pressurized gas bubbles;
- low diffusion capacity – slow migration of cations, anions and organic molecules;
- cation sorption capacity – ability to sorb and delay migration of cations;
- expandability – ability to expand and establish tight contact with the surroundings;
- suction – suction determines the rate with which initially unsaturated clay barriers hydrate.

Ductility
Clays in general and smectites in particular can undergo large strain without losing coherence, a practical measure being the Atterberg limits [1].

Hydraulic conductivity
The comprehensive research performed internationally for finding suitable clay materials for isolating canisters with highly radioactive waste has shown that sodium montmorillonite in dense form has the lowest hydraulic conductivity of all clays. Clays rich in mixed-layer minerals can stand less strain than montmorillonitic clays but their hydraulic conductivity is less influenced by high salt contents. Table 4.2 shows data for montmorillonite-rich MX-80 clay and Friedland Ton with montmorillonite/muscovite mixed-layer minerals as major component.

The table shows that the clay rich in expandable minerals (MX-80) is much tighter than the mixed-layer clay at low densities but that the difference is smaller for high densities. For higher densities the Friedland Ton appears to be less affected by high porewater salinity than MX-80, which is explained by the lower microstructural sensitivity of mixed-layer clays. Still, the montmorillonite-rich clay

Table 4.2: Hydraulic conductivity in m/s of MX-80 and Friedland Ton percolated by high salinity percolates [2].

	Density at saturation (kg/m^3)			
Percolate %	1600	1800	2000	2050
MX-80				
0.5 NaCl	7×10^{-12}	2×10^{-12}	2×10^{-13}	6×10^{-14}
3.5 NaCl	–	10^{-11}	10^{-12}	10^{-13}
0.5 $CaCl_2$	10^{-10}	2×10^{-12}	2×10^{-13}	9×10^{-14}
3.5 $CaCl_2$	10^{-9}	3×10^{-12}	3×10^{-13}	9×10^{-14}
Friedland Ton				
2 NaCl	2×10^{-8}	–	–	–
10 NaCl	5×10^{-8}	10^{-9}	5×10^{-11}	2×10^{-11}
3.5 $CaCl_2$	10^{-9}	6×10^{-10}	2×10^{-11}	–
10 $CaCl_2$	–	5×10^{-9}	3×10^{-11}	2×10^{-11}

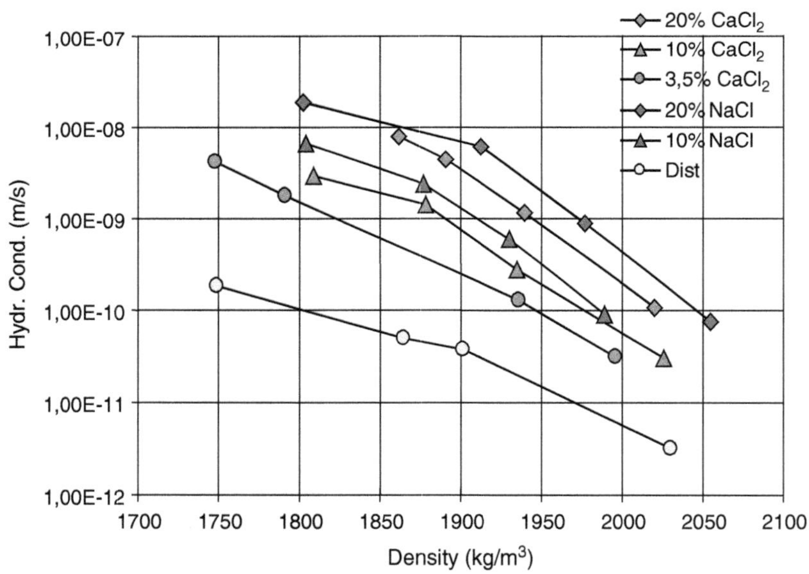

Figure 4.3: Hydraulic conductivity tests on saturated samples of Friedland Ton [2].

is superior at all densities. The influence of different salt contents on the hydraulic conductivity of Friedland Ton is shown in Fig. 4.3.

This diagram demonstrates that the conductivity is remarkably low even for extreme concentrations of cations in the porewater when the density exceeds about 2000 kg/m^3.

Table 4.3: Effective diffusion coefficient D_e for elements migrating in MX-80 bentonite with a density at saturation of about 2000 kg/m^3 [1, 2].

Species	D_e (m^2/s)
^{14}C	10^{-10}
^{129}I	2×10^{-12}
^{90}Sr	2×10^{-8}
^{137}Cs	2×10^{-9}
^{22}Na	10^{-9}
^{238}Pu	10^{-10}
^{243}Am	10^{-10}

Gas penetrability

Permeation of gas through clay takes place when a threshold pressure – the critical gas pressure – is exceeded. Once gas has made its way through the clay and further out through even more permeable geological units it will flow at a rate that is more dependent on the gas production rate than on the gas conductivity. The most important parameter is therefore the gas pressure that can initiate penetration of gas through the buffer clay. This pressure is on the same order of magnitude as the swelling pressure, which is higher for montmorillonite-rich clay than for clay consisting of mixed-layer minerals [1].

Ion diffusion capacity

In practice, the ion transport capacity can be evaluated from lab experiments by applying Fick's law and relevant values of the coefficient of 'effective' diffusion, D_e, for any density. Table 4.3 gives typical literature-derived data on this parameter for montmorillonite-rich clay with a density at saturation of 2000 kg/m^3. For lower densities the diffusion coefficient is higher; it is estimated to be one hundred times higher for a density of 1600 kg/m^3. For Friedland Ton it is assumed to be ten times higher than for MX-80 clay at densities exceeding about 1600 kg/m^3.

Cation sorption capacity

The lattice charge deficit gives the smectites an ability to adsorb and exchange charged cations and inorganic and organic molecules as manifested by the cation exchange capacity. Under a given set of conditions different cations are not equally replaceable and do not have the same replacing power. In principle, the following law of replacement power applies [1, 2] : Li < Na < K < Ca < Mg < NH_4.

In general Ca^{2+} will more easily replace Na^+ than Na^+ will replace Ca^{2+}, which is in turn easily replaced by Mg or NH_4. However, the replaceability varies depending on a number of factors, primarily the ion concentration, as demonstrated by the fact that the replacement of Ca^{2+} and Mg^{2+} by Na^+ in montmorillonite increases as the concentration of Na^+ in the solution increases. Na^+ vs. Ca^{2+}

represents a particularly important case of competition. Thus, it is a well known fact that as the amount of exchangeable calcium on the clay mineral becomes less it becomes more difficult for it to be released. Sodium, on the other hand, tends to become easier to release as the degree of saturation with sodium ions becomes less.

The matter is complex, as indicated by the fact that the cation exchange is affected by the nature of the anions in replacing solutions. This is illustrated by the fact that replacement of Na^+ by Ca^{2+} in montmorillonite depends on whether the electrolyte dissolved is calcium hydroxide or calcium sulphate. However, what affects the replaceability of cations most of all is the valence. The higher the valence of the cation, the higher its replacing power, with the exception of hydrogen, which behaves like a divalent or trivalent ion. For ions of the same valence, the replacing power increases with the size of the ion, meaning that smaller ions are less tightly held than the larger ions.

Potassium is an exception, which is explained by the fact that its ionic diameter 2.66 Å is about the same as the diameter of the cavity in the oxygen layer, so that the potassium ion can just fit into one of these cavities. As a consequence, the potassium ion is rather difficult to replace. For other cations it is the size of the hydrated ion, rather than the size of the non-hydrated one, that controls the replaceability. Thus, it appears that for ions of equal valence, those which are least hydrated have the greatest energy of replacement and are the most difficult to displace when present upon the clay. Li, although being a very small ion, is considered to be strongly hydrated and, therefore, to have a very large hydrated size. The low replacing power of Li^+ and its ready replaceability can be taken as a consequence of the large hydrated size but there are in fact indications that Li^+ and Na^+ are only weakly hydrated in interlamellar positions.

Typical cation exchange data of the most common smectite species are given in Table 4.4. For comparison, Friedland Ton has a cation exchange capacity of 40 CEC, meq/100 g.

Expandability

The expandability of smectitic clay is a most important property since it provides a self-sealing potential and makes sure that tight contact with the surrounding

Table 4.4: Typical CEC ranges for some important smectites [2, 3].

Species	CEC (meq/100 g)
Montmorillonite	80–150
Beidellite	80–135
Nontronite	60–120
Saponite	70–85
Mixed-layer minerals	30–50

rock is maintained. The major influencing factors are the density, porewater salinity and type of adsorbed cation.

Swelling pressure data for montmorillonite-rich clay (MX-80) and Friedland Ton saturated with low-electrolyte water are collected in Table 4.5. It demonstrates that montmorillonite-rich clay in sodium form has a swelling potential even at very low densities, while Friedland Ton does not. However, Fig. 4.4 shows that the latter is remarkably insensitive to very high salt contents at densities exceeding about 2000 kg/m^3.

Suction

The implication of the 'matric' and 'osmotic' forces that operate in clay/water systems is that suction in the porewater phase prevails when the degree of

Table 4.5: Swelling pressure (p_s) in MPa of a number of tested buffer materials saturated with distilled water [2].

Density at saturation (kg/m^3)	1300	1500	1700	1800	1900	2000	2100
MX-80 (MPa)	0.06	0.2	0.4	0.8–0.9	1.4	4–5	10–12
Friedland (MPa)	–	–	0.05	0.1	0.3	0.8–1	2–2.5

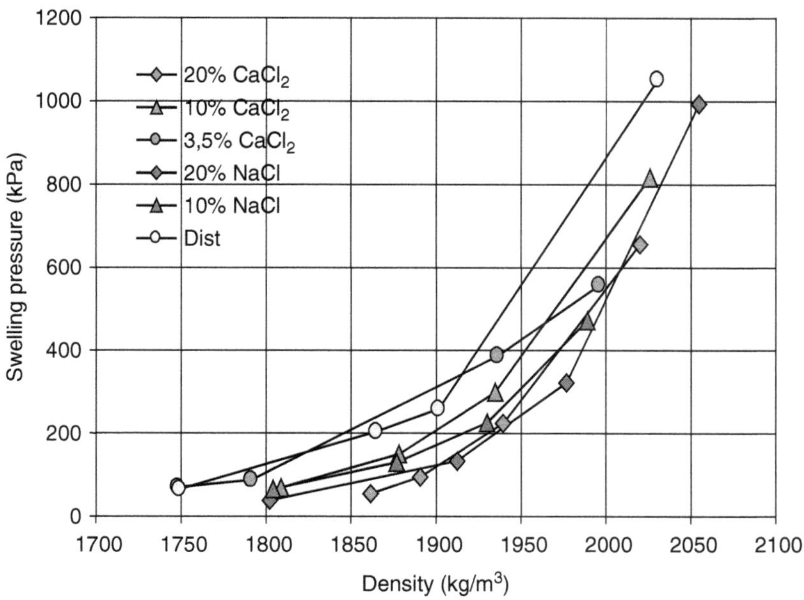

Figure 4.4: Swelling pressure tests on saturated samples of Friedland Ton [2].

Table 4.6: Example of suction as a function of dry density and degree of water saturation (S_r) of MX-80 clay.

S_r (%)	Suction (MPa) MX-80
30	100
40	70
50	50
70	20
90	10
95	8
99	1

saturation is below 100%. The suction, which is largely independent of the density, is insignificant when the degree of water saturation is higher than 95% (Table 4.6).

4.2 Methods for constructing engineered barriers in underground mines

4.2.1 Materials

Smectitic clays are proposed for isolation of solid and solidified waste, which are represented by alkaline and mercury batteries and clay-stabilized liquid pesticides, respectively. They form the main engineered barrier but need confinement in the form of concrete plugs placed strategically for sealing off drifts and rooms. Cement-based barriers will hence be placed in contact with waste-embedding clay, which makes the issue of chemical interaction of the two materials very important.

4.2.2 Preparation and application of smectite clay barriers

Clay materials like bentonites are usually treated in various ways before being packed and delivered. In a few cases the raw material is so uniform and rich in smectite that only simple drying in the sun of the excavated material with subsequent grinding has to be made at the quarries and plants. For Friedland Ton, which has sodium as the dominant adsorbed cation in nature, no other treatment than drying and grinding is required, while for Ca-bentonites mixing of sodium carbonate in powder form with the moist clay is required for bringing them to the desired sodium state.

The clay powder is commonly stored in silos or in stockpiles protected from rain and it is packed in 25–50 kg paper bags or 1000 kg big-bags for delivery. Bulk transport is often used but it is important to realize that this usually increases the water content, which may reach values as high as 18% for smectite-rich clay.

4.2.2.1 Compaction of blocks

Preparation of blocks under high pressure has been made and found possible without problems for water contents of the granules ranging between 10–17% for smectite-rich material and 7–12% for clays of Friedland Ton type. Mixing of clay powder and solid waste in the form of alkaline and mercury batteries to a volume ratio of 1 : 1 to 1 : 4 has been found feasible. A compaction pressure of 100 MPa results in a dry clay density of up to 2000 kg/m^3 (Fig. 4.5), while compaction under 10–30 MPa yields a dry density of up to 1600 kg/m^3. Cylindrical blocks weighing up to 2000 kg can be prepared by uniaxial or isostatic compression requiring robot technique for placement. Cheaper preparation can be made by uniaxial compression of blocks of handable size weighing about 20 kg as applied by the Hoeganaes Co in Sweden.

Alkaline batteries break at compaction under more than 10 MPa by which liquid is squeezed out and injected in the clay powder. This exposes the clay to the

Figure 4.5: Example of clay block compacted under 100 MPa uniaxial pressure. The dry density is about 1900 kg/m^3 (Geodevelopment AB).

hazardous content earlier than if the batteries remain physically intact, which hence speaks in favour of applying moderate pressures. Completely solid mercury batteries remain intact for more than 100 MPa pressure but the question is whether such high compaction is required and, if so, what density is really required. If dry densities of about 1500 kg/m^3 are sufficient one can consider another technique that has been investigated, i.e. layerwise application of clay mixed with solid or solidified waste.

4.2.2.2 Layerwise application and compaction

Compaction of layers of clays and mixtures of clays and solid constituents to dry densities of at least 1500 kg/m^3 has frequently been made in the construction of top and bottom liners of waste landfills. The LowRiskDT Project comprised field experiments with three layers of 25 cm, the central one consisting of Friedland Ton mixed with dichlorvos pesticide and the other two being clay with no admixtures. The mass ratio of clay and waste was 4/1, which is very conservative since high clay contents make effective compaction difficult.

Dichlorvos was used in an isopropyl solution (10% DDVP concentration). For the bottom and upper layer Friedland clay noodles were produced with the LAKER-equipment of FrieTon GmbH. The material with pesticide was prepared by an EIRICH Vacuum-Intensive-Mixer, type RV 15 Vac with 1 m^3 filling volume 1 t and quick rotation (40 m/s) under protection gas (Fig. 4.6). The liquid was added during the granulation. The optimum fluid content of the solidified granulates for effective compaction was found to be in the interval 18–21%. The mixing time was 4 t/h but EIRICH mixers are available for up to 1000 t/h.

Figure 4.6: EIRICH Vacuum-Intensive-Mixer.

Compaction was made by use of a Komatsu D 41 P caterpillar tractor and a BOMAG 213 DM-2 (13.0 t) vibrating padfoot roller as well as a BOMAG 213 DH-3 static smooth roller. There were four runs of each layer. Sampling and determination of the dry density produced the data in Table 4.7.

Figures 4.7–4.9 show field compaction (photos by DURTEC, Germany).

Table 4.7: Data from field compaction test.

Layers	Dry density (kg/m^3)	Water content (weight %)	Hydraulic conductivity (m/s)
First layer	1500	29	1.9×10^{-10}
Second layer	1550	26	2.1×10^{-10}
Third layer	1510	29	1.1×10^{-10}

Figure 4.7: Clay noodles of the first layer before compaction.

Figure 4.8: Compaction by vibrating pad-foot BOMAG-roller.

Figure 4.9: Homogeneous nature of compacted Friedland Ton.

4.3 Maturation of smectite clay barriers

4.3.1 Background

The clay embedment of chemical waste is initially saturated with water to significantly less than 100% and complete saturation may take a long time. The driving force for saturation is the hydration potential of the unsaturated clay. It causes tension by several tens of MPa at low water contents and drops to zero at complete saturation under constant volume conditions.

Current modelling of the rate of fluid saturation follows one of two principles: (1) suction in the buffer drives in water from the surroundings, i.e. the confining rock, at a rate that is determined by the transient hydraulic conductivity of the buffer and by the successively dropping suction potential. Expansion/consolidation and temperature impact on the maturing buffer are included. (2) The difference in water content of the freshly applied unsaturated clay and the fully wetted clay in contact with the rock causes diffusive water migration. The first principle cannot be quite adequately modelled at present while the second, being preferable from a scientific point of view, has the disadvantage that practically useful diffusion coefficients are not known with great accuracy. However, it is taken as a basis of modelling the maturation of clay barriers in this chapter.

4.3.2 Clay microstructure

Compaction of bentonite granules brings them into tight contact leaving voids of varying size in between. For low densities the voids make up continuous

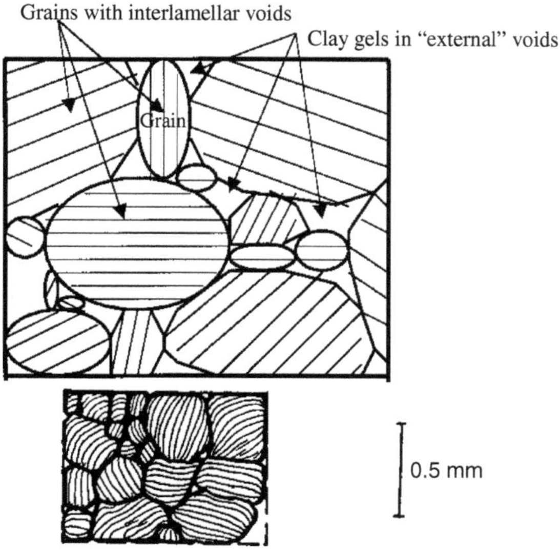

Figure 4.10: Particle arrangement in compacted clay powder. Upper: low dry density with voids between granules being up to several tens of micrometres wide. Lower: high density by compaction under several tens of MPa. The grains contain numerous small, isolated voids.

channels with constrictions while the voids are isolated in dense clays on all scales (Fig. 4.10).

4.3.3 Hydration

4.3.3.1 Mechanisms

The rate of hydration of the waste-embedding clay is of fundamental importance to the performance of a mine repository since significant dissolution and transport of released toxic elements is not initiated until the clay is largely fluid-saturated.

One can distinguish between three major cases that represent different hydraulic boundary conditions:

- contact of dry clay with water vapour;
- contact of dry clay with non-pressurized liquid water;
- contact of dry clay with pressurized liquid water.

In the first case, the process is dominated by migration of water molecules into open voids and get sorbed on exposed mineral surfaces from where they migrate into the interlamellar space, which has the highest hydration potential. Since the migration takes place along surfaces with differences in hydration potential (water content) as driving force, the entire process is one of diffusion. The saturation process is sufficiently slow to let entrapped air be dissolved and diffuse from the clay without delaying the saturation process.

In the second case, water is sucked up by capillary forces in the open channels from which water molecules migrate into finer voids and further along mineral surfaces by diffusion into the interlamellar space. Since the clay matrix expands thereby, the large channels become closed rather early and the hydration is then controlled by diffusion along mineral surfaces at the wetting front. The matrix and osmotic potentials are similar to those in Case 1 but the hydration rate is believed to be somewhat higher because the larger channels are filled more quickly. The degree of saturation will probably never be 100%.

In the third case, which will prevail after some tens to hundred years in a mine repository, water is pressed into the largest open channels and moves quickly and deeply into the clay matrix, particularly when the pressure and electrolyte contents are high. The penetrating water displaces air and compresses the unsaturated matrix. Since a considerable fraction of the voids become water-filled quickly, the average degree of saturation is raised very early and the dominating process is hence diffusive redistributing water from the larger channels into the clay matrix, associated with diffusive particle movements. The ultimate degree of saturation will be 100%.

4.3.3.2 Rate of hydration

The diffusive character of the hydration process is illustrated by a number of laboratory and field tests. Fig. 4.11 shows a comparison of recorded and predicted water saturation of a 50 mm long confined cylinder of highly compressed

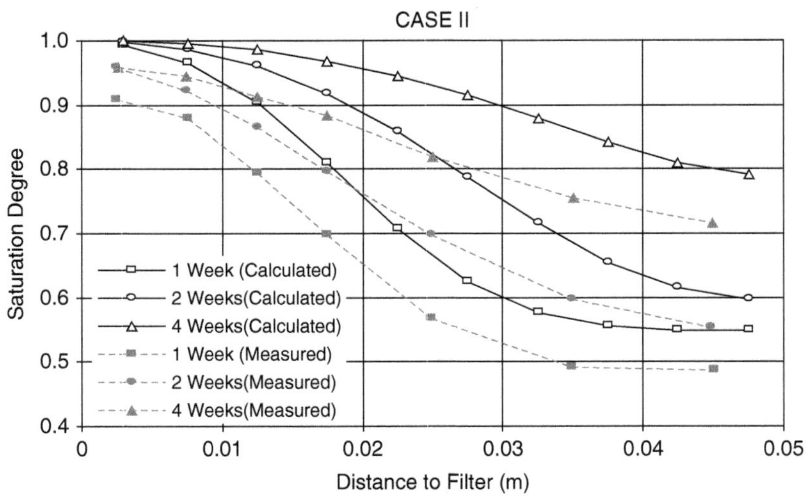

Figure 4.11: Predicted (black curves) and recorded rate of water saturation of 50 mm long cylinder of highly compressed MX-80 clay (after A.Ledesma).

smectite-rich clay (MX-80) with a dry density of 1900 kg/m^3 with about 50% initial degree of water saturation. The prediction was made by assuming that the wetting is caused by flow driven by the suction in the clay matrix. The predicted rate of hydration is significantly higher than the recorded, which corresponds to a diffusion coefficient of 3×10^{-10} m^2/s.

Extrapolation of the wetting process shows that the time for saturation of a dense smectite-rich clay like block containing batteries can be extreme. Thus, diffusive fluid saturation of a 1 m thick very dense smectite-rich clay barrier will take thousands of years. For less smectite-rich clay like the Friedland Ton the theoretical time for fluid saturation is shorter because the diffusion coefficient is higher – about 10^{-9} m^2/s for high densities – but still very significant. Thus, a 1 m thick barrier of this kind of clay with an original degree of saturation of 30–40% clay and a dry density of 1500 kg/m^2, corresponding to a density at fluid saturation of 1950 kg/m^3, will not be fully saturated until after 4000 years (Fig. 4.12).

Impact of water pressure
Experiments with smectite-rich clay (MX-80) compressed under 30 MPa pressure giving it a dry density of kg/m^3 and a degree of water saturation of 50% have shown that water at a pressure of 0.6 MPa enters the clay by a few centimetres in a few minutes and then stops [4]. This suggests that continuous systems of larger voids between granules are quickly filled to a certain shallow depth but that expansion of the clay matrix surrounding these channels closes them and prevents further penetration by flow. For less expandable clay like the Friedland Ton a somewhat but not significantly quicker water penetration is expected.

Importance of the ability of the rock to give off water to the waste-embedding clay
The average hydraulic conductivity of a crystalline rock mass is in the interval 10^{-8}–10^{-11} m/s, while it is appreciably lower for argillaceous rock and salt rock.

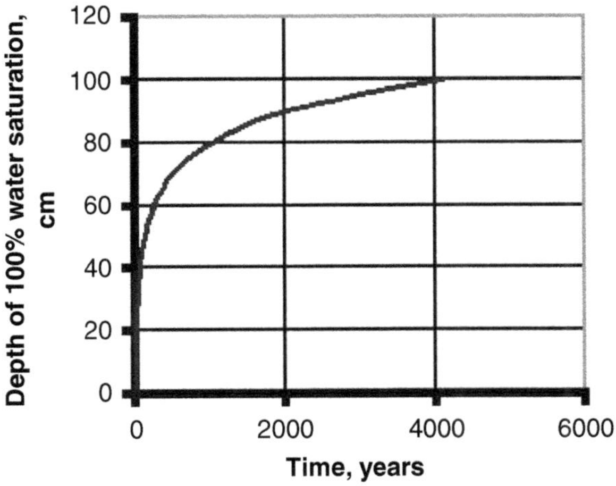

Figure 4.12: Development of fluid saturation for $D = 10^{-9}$ m^2/s.

This parameter determines the groundwater flow driven by hydraulic gradients and it is clear that if the conductivity is very low it can further delay fluid saturation of clay barriers and hence release of toxic elements from the waste. The matter has been considered by theoretical modelling of the hydraulic interaction of rock and clay-embedded waste assuming the diffusion coefficient for fluid saturation of the clay and different hydraulic conductivities for the rock. The modelling was made in the fashion described below.

Diffusion is controlled by:

$$\frac{\partial C}{\partial t} = D\nabla \cdot (\nabla C), \tag{4.1}$$

where C is the degree of water saturation (initially 0.5)

Flow in rock:

$$S\frac{\partial \phi}{\partial t} = K\nabla \cdot (\nabla \phi), \tag{4.2}$$

where K is the conductivity and ϕ the pressure

Porosity for the clay is defined as:

$$P = \frac{V_{\text{por}}}{V_{\text{tot}}} \Rightarrow V_{\text{por}} = PV_{\text{tot}}. \tag{4.3}$$

One has

$$C(r = r_1) = \int \frac{q_{\text{w}}(r = r_1)}{V_{\text{por}}} \text{d}t, \tag{4.4}$$

where $q_{\text{w}}(r = r_1)$ is the flow for $r = r_1$ i.e. the inner radius of the rock, and $q_{\text{w}}(r = r_1)$ is the well formula with ϕ_1 and ϕ_2 representing the pressure potentials at distances r_1 an r_2

$$q_{\text{w}} = K2\pi b(\phi_2 - \phi_1) / \ln\left(\frac{r_2}{r_1}\right). \tag{4.5}$$

For *underpressure* at $r = r_1$ flow takes place across the boundary $r = r_1$. Mass balance for r_1 gives for $r = r_1$:

$$C(r = r_1) = \int \frac{K2\pi b(\phi_2 - \phi_1(t)) / \ln(r_2/r_1)}{PV_{\text{tot}}} \text{d}t. \tag{4.6}$$

Taking the wetting of the clay to be diffusive with $D = 3 \times 10^{-10}$ m²/s and the bulk rock conductivity $K = 10^{-10}$ m/s and the initial degree of saturation to be 0.5 one gets the diagram in Fig. 4.13. Comparing the rate of saturation with a case with unlimited access to water for saturation of the clay one finds that the conductivity of the rock is sufficient for supplying the clay with water without delay.

Considering now a tighter rock with $K = 10^{-10}$ m/s hosting clay with $D = 3 \times 10^{-10}$ m^2/s and an initial degree of saturation of 0.5, one gets the diagram

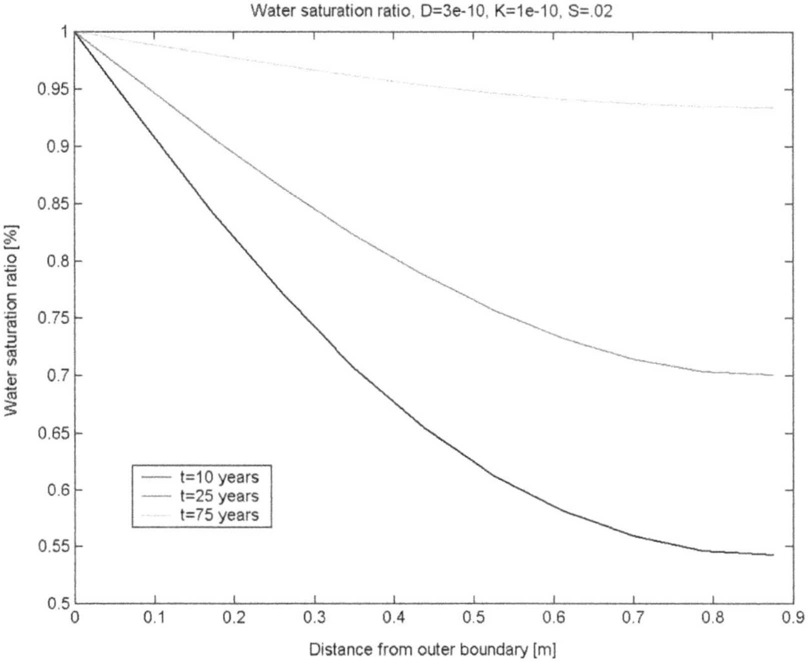

Figure 4.13: Water saturation stages in smectite-rich clay barrier in rock with an average hydraulic conductivity of 10^{-10} m/s.

in Fig. 4.14. Comparing the rate of saturation with a case with unlimited access to water for saturation of the clay one finds that the conductivity of the rock is sufficient low to delay wetting of the clay. Considering finally rock with $K = 10^{-12}$ m/s with the same clay as in the previous cases Fig. 4.15 shows that the clay sucks water from the rock, which further delays saturation. It is important to realize that until 90% saturation has been reached there is no free water available in the clay for chemical reactions with embedded waste.

The diagram in Fig. 4.15 underlines the advantage of disposing waste in argillaceous and salt rock, which usually have hydraulic conductivities that are lower than 10^{-12} m/s. For a conductivity of less than this figure comprehensive dissolution of waste in the fully water-saturated clay and concomitant release of hazardous elements will not take place in the first tens of thousands of years after emplacement. It is important to realize that the role of the excavation-disturbed zone (EDZ) with its high hydraulic conductivity is very limited if it can be cut-off so that isolated, stagnant hydraulic regimes are created. This can be made without difficulties by constructing clay plugs that extend into the surrounding rock at strategic sites, i.e. away from fracture zones. The clay plugs need to be supported by concrete bulwarks, which endangers the long-term chemical stability of cement and clay.

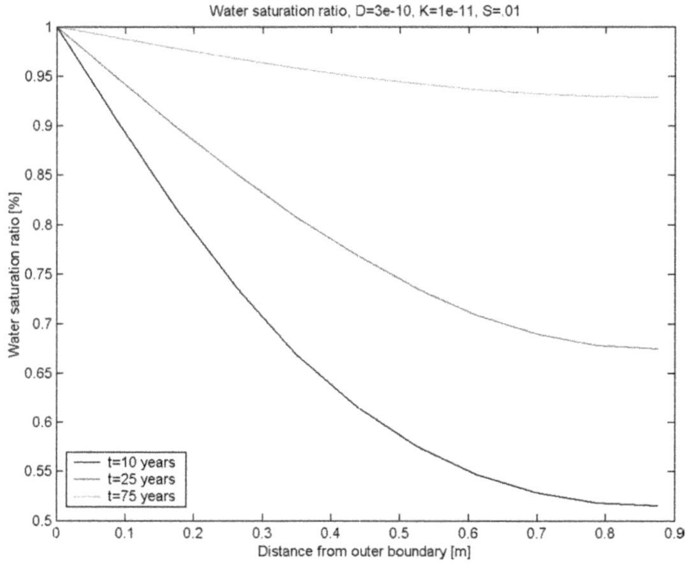

Figure 4.14: Water saturation stages in smectite-rich clay barrier in rock with an average hydraulic conductivity of 10^{-11} m/s.

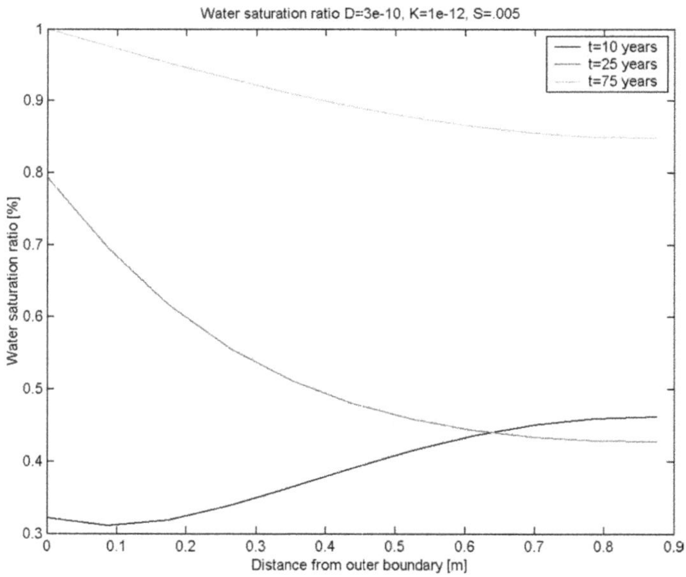

Figure 4.15: Water saturation stages in smectite-rich clay barrier in rock with an average hydraulic conductivity of 10^{-12} m/s.

4.4 The source term

4.4.1 Definitions

The source term is the concentration of hazardous chemical elements that are released from the waste. Assuming that the mineral phase is chemically stable the source term will be defined as the concentration at the contact between the clay and the surrounding rock, i.e. the EDZ. The definition of the ion concentration representing the source term depends on the following factors: For describing the source term the following steps must be taken:

- define the composition of the clay-embedded waste;
- define the clay;
- identify hazardous species;
- define the time for complete fluid saturation of the clay, i.e. the time when significant dissolution of the waste starts;
- define the rate of dissolution of waste, i.e. the rate of release of hazardous species;
- define the rate of migration of hazardous species from the waste to the EDZ;
- define the concentration of hazardous species at the EDZ boundary, i.e. the 'source term'.

In practice, these factors imply that the source term is practically nil in the first few thousand years after waste application and that it increases very slowly if the dissolution of the waste and the migration of released ions are slow. These matters have been investigated in several series of laboratory experiments, which are summarized here.

4.4.2 Tests

4.4.2.1 Alkaline batteries in Friedland Ton

A major test of mixing physically intact alkali batteries in Friedland Ton powder and compacting the mixture under 30 MPa pressure in a 100 mm cell, yielded a dry density of 1550 kg/m^3 with 30% degree of fluid saturation. Water saturation was made with 3.5% $CaCl_2$ and 20% NaCl solution for simulating extreme conditions in deep mines. It was found that the strong compression of the mixture of clay and non-corroded alkaline batteries had squeezed out electrolytes from the batteries and that Ca, Zn and Ni had entered the surrounding clay but only to a very small distance (Fig. 4.16). Taking the migration of Zn that had migrated farthest after 10 months, as caused by diffusion, it yielded the *D*-coefficient 10^{-13} m^2/s, which is much lower than what is typical for polyvalent cations (Fig. 4.17). The reason for the discrepancy is believed to be complexation and fixation of the reaction products.

Figure 4.16: Split clay sample with alkaline battery from which electrolytes were squeezed out forming halos.

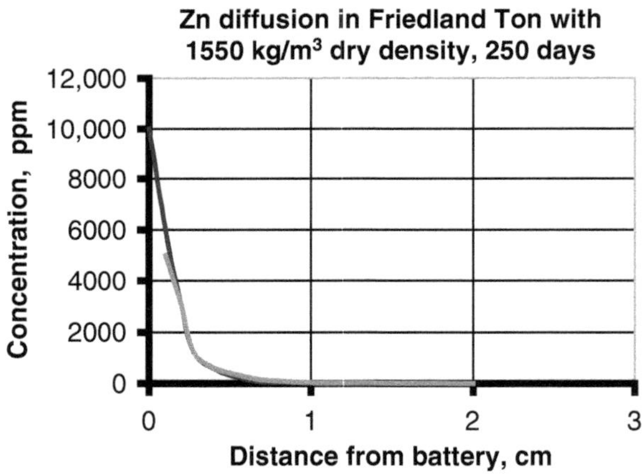

Figure 4.17: Zn concentration in the clay after 250 days. The curve reaching up to 10,000 ppm is the diffusion profile for $D = 10^{-13}$ m^2/s, which coincides with measured data.

Additional tests were made with strongly corroded Hg batteries with 30 mm diameter and 60 mm length (Fig. 4.18), and uncorroded button-type Hg batteries with 12 mm diameter and 4 mm height. Compaction of the clay was made to about 1 cm height from the lower filter in the cell and the batteries were then placed on this level and covered with clay powder that was compacted to a relatively

Figure 4.18: Strongly corroded Hg batteries (Photo: R. Sjöblom).

low density. In both tests saturation of the clay was made with electrolyte-free water that was also used for percolation.

Rather soft Friedland Ton was used in these later tests for finding out what the effect would be of embedding the waste in sandwiched layers of clay/waste mixtures and pure clay applied in big rooms in a mine. The dry density was 1400 kg/m^3, which corresponds to a density at water saturation of 1900 kg/m^3. It can easily be reached by compaction with vibrating rollers and plates, which means that this test series represents the cheapest possible way of applying the clay mixed with batteries and compacted on site.

In the test with initially intact batteries the water content distribution 3 and 10 months after start is estimated to have been as shown in Fig. 4.19. The wetting process is of diffusion type with the coefficient 10^{-9} m^2/s.

4.4.2.2 Chemical interaction of clay and corroded batteries

The cell was opened 10 months after the test start and small samples taken at different distances from the batteries, which hade given off ammonium in gaseous form that had delayed water saturation. The estimated rate of diffusion of the most mobile toxic element, Zn, is in the range of 10^{-10} till 10^{-9} m^2/s disregarding from retardation by complexation which would reduce this value.

The elements Na, K, Ca, Cu, As and Cd did not show any dependence of the distance from the batteries at all, they are all present as natural constituents in the Friedland Ton. One finds a certain minor concentration gradient for Cr and Ni but it can have been caused by variations in the natural clay that contains these

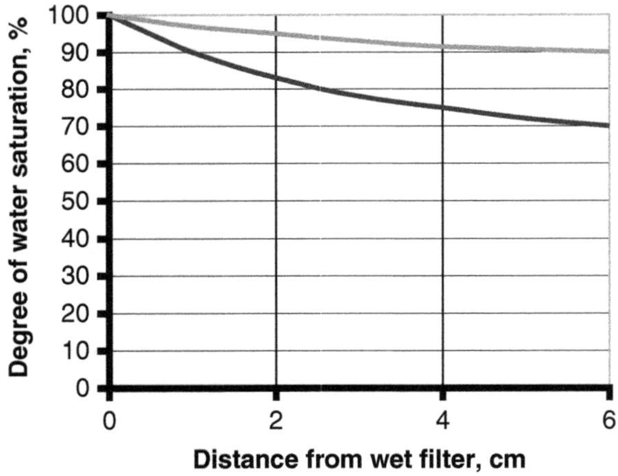

Figure 4.19: Graph showing the rate of wetting. The lower curve represents 3 months after start and the upper 10 months after start.

elements as well. The only obvious concentration gradient is for Zn and Hg, which have very low concentrations in the natural clay. The Hg concentration in the tested samples is insignificant and as for alkaline batteries the only important migration of toxic species is for Zn, which can therefore be taken as 'reference element' in the definition of the source term for batteries.

4.4.2.3 Chemical interaction of clay and uncorroded Hg batteries

The last test series, i.e. the 18 month-long experiments with uncorroded Hg batteries in clay with moderate density, showed that no ion release at all from the batteries had taken place. Figure 4.20 illustrates the typical appearance of the sectioned clay in conjunction with sampling for analyses. The clay was perfectly uniform with no colour changes or other indications of chemical reactions. The originally shiny batteries had retained their appearance except for some slight greyness indicating very slight chemical interaction with the clay minerals and porewater. The corrosion depth cannot have exceeded 10 μm as concluded from light microscopy.

As expected, the analysis showed that massive ion exchange from initially sorbed sodium to calcium had taken place at the saturation and percolation of calcium chloride. The concentrations of Ni and Sr and, naturally, Hg were very low indicating that the batteries had not undergone measurable dissolution.

4.4.2.4 Other hazardous waste

Organic pesticides were also considered in the LowRiskDT Project. They are in liquid form and the first step in storing them underground is to make solidification, which can be successfully made by mixing the liquid pesticide with Friedland Ton

Figure 4.20: Sectioned clay sample. The shiny surface of the batteries indicate almost no corrosion after percolation with distilled water for 12 months and subsequent percolation for 6 months with 3.5% $CaCl_2$ solution.

to the weight proportions 1 : 3. The pesticide Dichlorvos was found to be sorbed in the interlamellar space and on the base surface of the clay crystallites and gave about the same physical bulk properties of the clay as by saturating it with $CaCl_2$ solution. The study led to the conclusion that Zn and the principal organic molecule of the active component DCA of Dichlorvos are sorbed in the same fashion as the calcium ion. In turn, this suggests that migration of these species takes place according to the same diffusion process and coefficient.

The initial concentration of Dichlorvos in the repository is assumed to be 10,000 ppm and since early conversion to DCA is assumed the concentration of DCA inside the repository is equal to 10,000 ppm.

4.5 Basis for modelling transport of hazardous elements from the waste

4.5.1 General

4.5.1.1 Definition of the source term

The most suitable way of defining the source term is to express the ion concentration at the contact between the clay and the rock as a function of time since it is determined by the wetting rate of the clay, the rate of released waste into the clay, and the rate of migration of released ions. In practice, sufficient chemical data are not available for deriving mathematical expressions of the source term and it is therefore practical to define it in a simple and safe way. One way is to assume the worst case scenario according to which an element of waste dissolves instantly when the surrounding clay is fully water saturated.

As one practical example we consider the source term for disposal of alkaline and Hg batteries. The basic parameter is the clay/waste mass ratio and for the cases of alkaline and Hg batteries Zn is the major hazardous chemical species, contributing with up to 25% of the battery mass. The mass ratio of solid (dry) clay and batteries depends on the dry density of the clay that can range between 1400 and 1600 kg/m^3. Batteries weigh about 2500 kg/m^3, which hence means that the mass ratio can be in the interval 1 : 1 to 1 : 20.

Migration rate of released hazardous ions

The major rate-controlling parameter is the rate of wetting of the clay. As assumed in the project, Zn starts migrating towards the rock as soon as the particular clay volume is water saturated. This rate is expressed as a diffusion coefficient that can be as low as 10^{-13} m^2/s for very densely compressed clay and up to 10^{-9} m^2/s for clay densities that can be achieved by using ordinary field compaction techniques.

Generalization

Applying the above concept and the diffusion coefficient 10^{-9} m^2/s for both wetting and diffusion of the hazardous elements one can identify the following stages, omitting the possible impact of water pressure on the clay wetting rate:

- In 0–2 years there is simultaneous wetting of the clay in the most shallow 20 cm thick clay/waste annulus (that becomes largely water-saturated, in this period). If there are hazardous elements in this layer they will start migrating.
- In 2–20 years there is simultaneous wetting of another 10 cm of the clay/waste mass and migration of the hazardous elements from the wetting front that is finally 30 cm from the EDZ.
- In 20–200 years there is simultaneous wetting of another 20 cm of the clay/waste mass and migration of Zn from the wetting front that is 50 cm from the EDZ.
- In 200–4000 years there is simultaneous wetting of another 50 cm of the clay/waste mass and migration of Zn from the wetting front that is finally 100 cm from the EDZ.
- In 4000–100,000 years there is simultaneous wetting of another 150 cm of the clay/waste mass and migration of hazardous elements from the wetting front that is ultimately located in the centre of the drift.

In all these stages the source term may be considered as the average concentration of hazardous elements at the clay/EDZ contact.

One realizes that if the waste mass in the big room is surrounded by a 100 cm 'liner' of Friedland Ton with the assumed density (see Fig. 4.21), practically important wetting of the volume of clay-embedded waste will not commence

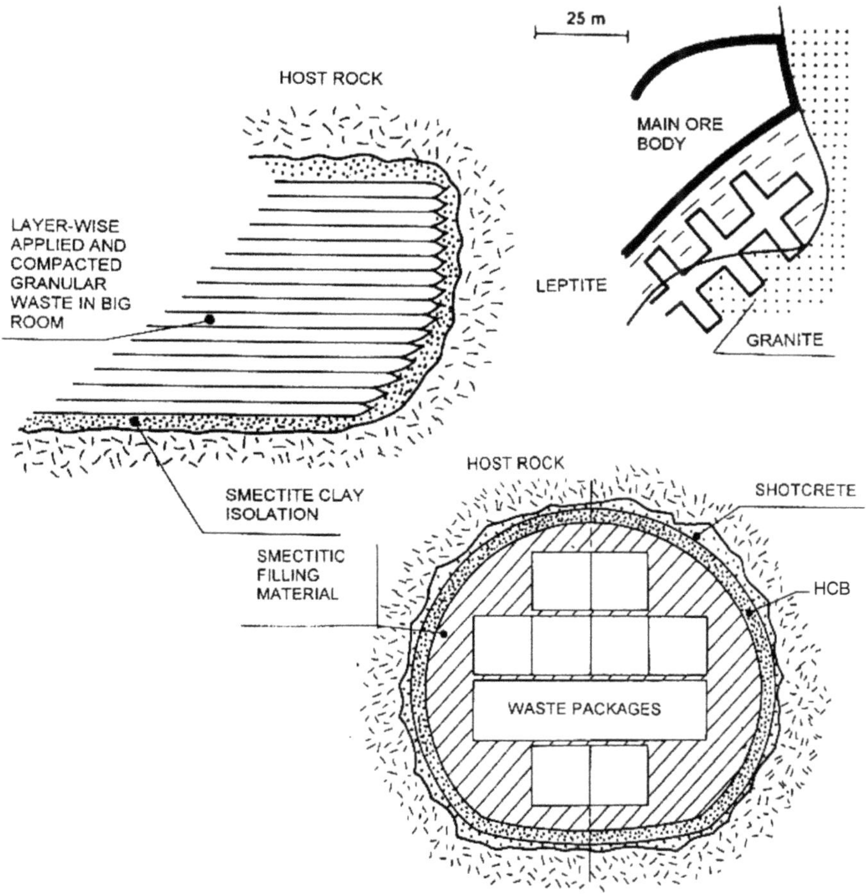

Figure 4.21: Example of possible repository concept with waste-free clay zone adjacent to the rock and surrounding waste/clay mixtures. HCB denotes a liner of highly compacted bentonite.

until 4000 years after application, disregarding the impact of pressure-induced wetting.

4.5.2 Safety aspects

In addition to uncertainties in selecting representative diffusion coefficients, densities and geometrical data in applying the clay-based isolation principles described in this chapter there is also incomplete knowledge as to the detailed performance of the clay matrix when exposed to gas pressure and heat. Added to this is the question of longevity, which is particularly important and will be considered separately in the subsequent chapter. However, the very comprehensive research performed in the international work for finding safe methods for deep geological disposal of highly radioactive waste provides strong support to the techniques described here.

4.6 Long-term chemical stability of smectite

4.6.1 General

Smectites are sensitive to low and high pH and high temperature. Critical pH conditions can result from the waste or reactions of clay and waste, a well-known problem is the impact of high-pH cement on smectite.

4.6.2 Conversion of smectite to non-expandable minerals ('illitization')

The basic conversion process when smectite is transformed to non-expandable minerals, primarily illite, is [1]:

$$\text{Smectite} + Al^{3+} + K^{+} \text{ -------- } \text{Illite} + Si^{4+}.$$

The reaction is of Arrhenius type, meaning that the activation energy determines the degree of conversion as a function of time provided that access to potassium is not a limiting factor. Figure 4.22 illustrates the conversion rate for commonly accepted activation energies.

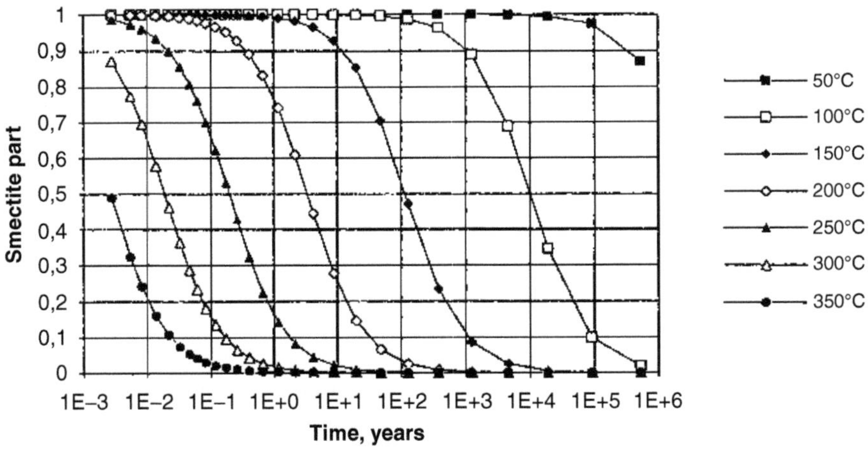

Figure 4.22: Diagram showing the expected conversion of smectite to illite for the activation energy 27 kcal/mole according to Pytte/Reynolds/Huang. 'Smectite part 1' represents 100% smectite, which drops to 97% in 10^5 years at 50°C and to 50% in 100 years at 150.

According to the current conversion model, clay-embedding waste in the pH range of 6–10 and temperatures lower than 50°C will consequently be largely intact even after one hundred thousand years.

4.6.3 Chemical interaction of smectite clay and cement

It was realized a number of years ago that Portland cement, which makes up some 30% of concrete, and smectite may interact chemically and that this may lead to a loss of the isolation potential of the clay and of the strength of the concrete. The ECOCLAY Project [5] comprised batch leaching tests with bentonite mixed with four synthetic cement porewater solutions and determination of chemical and microstructural changes at the cement/bentonite interface. Also, tests were made by compaction of MX-80 in a cell to 1400 and 1600 kg/m^3 dry density with subsequent drilling of a hole in the bentonite and filling it with fresh cement paste. This closed system was examined after 3–12 months with respect to chemical and microstructural changes. In a third test series a hardened cement disc was placed in contact with bentonite and the system percolated by granitic water. The quantity of dissolved smectite was found to depend on the water content. The average amount of dissolved smectite was 12 g of smectite per litre of pore solution, i.e. 1.2% under the closed conditions that prevailed.

Theoretical considerations [6] have indicated that zeolites like phillipsite and analcime should be reaction products and this has also been verified by tests showing that they can be formed in a few months in batch tests with $KOH/NaOH/Ca(OH)_2$ at 90°C and after several months at lower temperatures. Interstratified smectite/illite (S/I) with up to 15–20% illite was found when the solution contained much K^+. Uptake of Mg^{2+} in the montmorillonite crystal lattice yielded the smectite species saponite.

The most important conclusions from these earlier studies were:

1. high-alkali cement degrades quicker than low-alkali cement, which degrades by destruction of the CAH gel;
2. dissolved elements and water migrate from the fresh cement to the smectite clay in the first few hours;
3. the cement paste is dehydrated and its voids become wider; water moves from the clay to the dense cement matrix;
4. the dense cement paste fissures;
5. pH = 12.6 is a critical value for significant changes of the smectite component;
6. calcium migrates from the cement to the clay causing ion exchange and change in the microstructure of the clay by coagulating softer parts.

The ECOCLAY Project comprised testing of cements of Portland type and some low-pH cements and gave information on the type of reactions that can change smectite-rich clay and cement. Later investigations of the chemical interaction of Friedland Ton and two low-pH cements used the equipment shown in Fig. 4.23 [7].

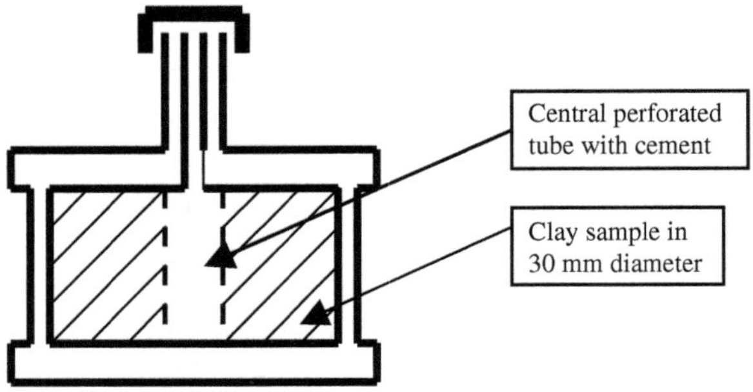

Figure 4.23: Cell for clay/cement–water experiments.

Table 4.8: Cement porewater composition in mg/l.

Type	Time (months)	Si	Al	Mg	Na	Ca	K	Fe	Cl	SO_4	pH
M	0	3.2	0.3	16.9	1260	30	107	0.2	410	1600	9.4
	2	5.8	0.9	51.0	2030	155	82	1.8	930	3440	7.8
	4	5.0	0.6	31	1450	69	99	1.0	381	1740	7.9
	5	6.2	0.6	22.7	1070	74	112	1.0	211	1100	7.9
E	0	1.9	2.5	19.2	1540	39	117	0.3	607	2070	8.1
	2	3.7	20.7	21.8	1890	113	96	6.1	946	2760	7.9
	4	3.5	35.7	27.9	1450	78	112	1.0	459	1630	8.0
	5	6.7	19.1	27.0	1130	219	98	5.3	228	1010	8.1

M, MERIT 5000; E, ELECTROLAND.

Clay powder, which was compacted in the cells to a dry density 1400 kg/m^3, surrounded a central filter tube for preventing clay particles to migrate into the cement–water solution that was contained in the tube. The solution was replaced at the end of a number of 1–2 month-long test periods and analysed with respect to pH and the concentration of important elements. After 5 months part of the clay samples were extracted and examined by the use of X-ray diffraction technique and electron microscopy, the rest of the clay samples were used for determination of the hydraulic conductivity. Replacement of the cement–water once per test period maintained the condition of high-concentration cement–water throughout the experiment.

The two low-pH cements, Swedish MERIT 5000 and Spanish ELECTROLAND, are recommended for construction purposes by the manufacturers. The first mentioned has 34% SiO_2, 13% Al_2O_3, 17% MgO and 31% CaO, while the other has 3% SiO_2, 41% Al_2O_3, 39% CaO and 16 Fe compounds. Cement porewater was prepared by mixing the respective cement powder with distilled water to w/c equal to 1.0 and separating the supernatant. Table 4.8 shows the initial composition of the cement solutions (0 months) and the composition after different period of time.

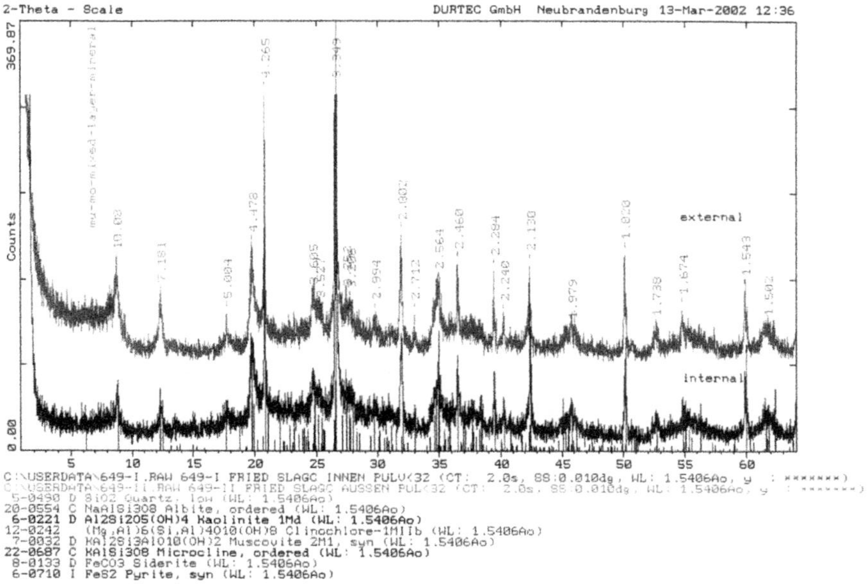

Figure 4.24: XRD diagram of Friedland clay far from the cement–water ('external') and close to it ('internal') after 5 months reaction with porewater of slag cement (Merit 5000).

The most important conclusions from the various analyses are [7]:

1. pH remained nearly constant in the solutions and ranged between 7.4 and 9.4, i.e. substantially lower than for Portland cement, hence suggesting much less attack on the clay minerals.
2. Some dissolution of the clay is manifested by the increased Si, Al, Mg and Fe contents in the solutions. The stronger loss of aluminium from the clay to the ELECTROLAND solution indicates a somewhat more extensive dissolution than for MERIT 5000.
3. The increase in iron in the solutions indicates that dissolution of the chlorite component of the clay may have dominated the degradation of silicate minerals in the clay.
4. The insignificant change in potassium content in the solutions suggests that illitization was absent or insignificant. This is also manifested by X-ray diffraction analyses (Fig. 4.24), which showed no differences between clay close to the filter with cement solution and far from it, nor from natural Friedland Ton.
5. The well-known microstructural constitution of Friedland Ton was preserved after exposure to cement–water as demonstrated by scanning microscopy equipped with a KEVEX-System for energy dispersive element analysis (Fig. 4.25). Thus, no zeolites were formed.

6. At the end of the cement–water treatment the hydraulic conductivity of the clay samples was determined and found to be almost exactly the same as for natural Friedland Ton.

The XRD analyses showed that no changes in the mineralogical composition had taken place but that cation exchange by uptake of sodium from the cement solutions had occurred. This is manifested by a change of the position of the broad (001) reflection of the muscovite–montmorillonite mixed-layer mineral from 12 to 14 Å in the MERIT 5000 case.

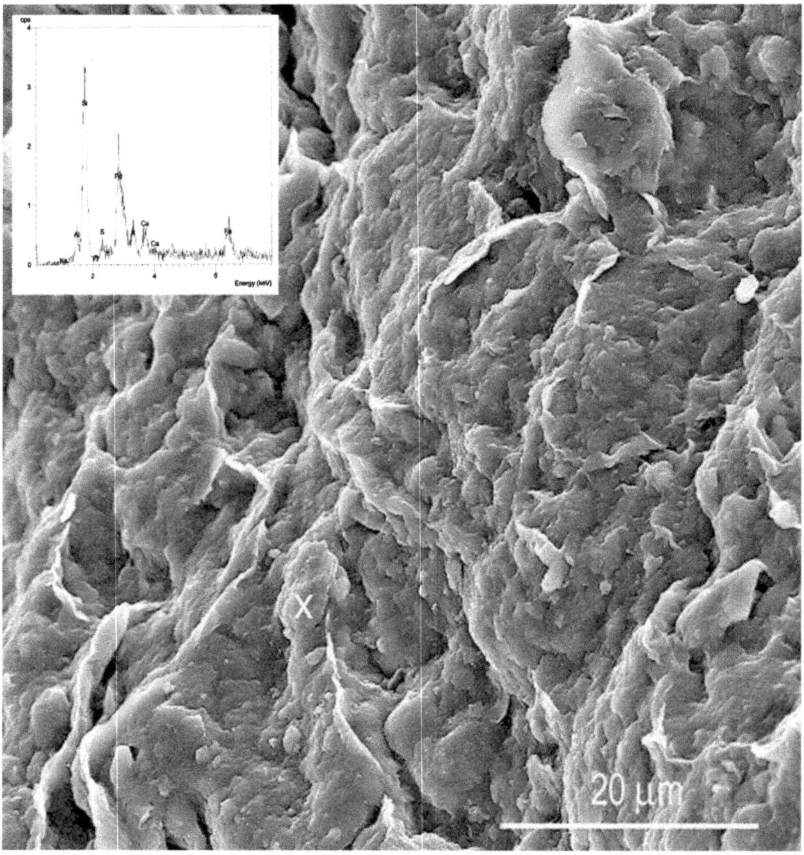

Figure 4.25: SEM micrographs of Friedland Ton reacted with Merit 5000 cement–water showing the typical microstructure of Friedland Ton with no minerals of zeolite type present. The element analysis shows a spectrum typical of natural Friedland Ton with some enhanced calcium content in the case of MERIT 5000. Microscopy by Dr J. Kasbohm, Greifswald University, Germany.

In conclusion, the answer to the important question of whether concrete bulwarks in a mine repository can seriously affect Friedland Ton is that they will not, provided that low-pH cement is used for the construction.

4.7 Cost estimates

One of the most important issues for assessing the applicability of the proposed techniques for isolation of chemical waste is cost and it has been in focus throughout the study. The outcome of the investigation of cost is reported here.

4.7.1 Disposal of batteries (mixed with clay powder and compacted to blocks)

The geometry of the room considered for disposal of compacted clay blocks containing batteries is assumed to be 50 × 50 × 20 m = 50,000 m^3. The volume ratio of clay and batteries is conservatively taken as 50/50. In practice the ratio can probably be reduced to 1/10. Two clay components are used:

- clay granulate used for filling the space between tunnel wall and compacted blocks (thickness approximately 1 m);
- clay blocks 240 × 240 × 190 mm.

Pos.	Performance	SP/m^3* (€)	SP/m^3 u.R.* (€)
1	Block preparation including clay powder mixing, pressing on selected press equipment	64.78	56.71
2	Placement of blocks in caverns	147.37	129.02
3	Filling of clay granulate	52.50	6.54
	Total	264.65	192.27

*Cost estimate excepting transport and intermediate storage; SP/m^3 – single price per m^3 compacted clay/battery blocks; TP – total price SP/m^3 u.R. – cost for room filling.

For block preparation one can use the following equipment:

No.	Machine/tool	Type	Investment cost
1	Press	HDP 800	ca. 1,50 Mio €
2	Block puzzle robot	VM 204 Robotec	35,500,00 €

4.7.2 Disposal of solidified pesticides (sandwiched clay and clay/waste layers)

The geometry of the room considered for disposal of solidified organic pesticides is assumed to be 50 × 50 × 20 m = 50,000 m³. The following build-up of layers utilizing two types of clay is assumed:

1. layer: clay noodles, compacted (*d* = 25 cm)
2. layer: granulated (1:4 liquid/clay) (*d* = 25 cm)
3. Large: clay noodles, compacted (*d* = 25 cm)
4. Large: granulated liquid waste (*d* = 25 cm)
   - clay granulate is used for filling 3 m space between tunnel wall and compacted blocks
   - volume of filled-in clay granulate = 17.088 m³.

Pos.	Performance	SP/m³* (€)	TP* (€)	SP/m³ u.R.* (€)
1	Preparation of 1. clay noodle layer (material delivery, ex-work, preparation of uncompacted layer, compaction and nivellement)	44.50	732,292.00	14.65
2	Preparation of 2. layer (delivery of clay powder, ex-work, mixing and granulation with ratio 1 : 4 (clay powder), preparation of uncompacted layer, compaction and nivellement	96.75	1,592,118.00	31.84
3	Filling up with clay granulate of the remained space room including Material delivery (ex-work Friedland) – granulate compaction	52.50	897,120.00	17.94
	Total		3,221,530.00	64.43

*Cost estimate excepting transport and intermediate storage; SP/m³ – single price per m³; TP – total price; SP/m³ u.R. – specific price for room filling.

For material preparation and placement of clay layers the price expected to be offered by contractors is as follows:

No.	Machine/tool	Type	Effective weight/useful volume	Investment costs (net)
1	Müllverdichter (vibratory roller)	CAT 816 F CAT 826 G	25 t 35 t	220,000,00 € 365,000,00 €
2	Glattmantelwalze (ordinary roller)	CS 683 E	19 t	100,000,00 €
3	Kettenbagger (chain excavator)	CAT 330	35 t	215,000,00 €
4	Dumper	CAT 740	22 m³ Muldeninhalt	235,000,00 €
5	Dozer	D 7 R	30 t	200,000,00 €
6	Wheel-loader (Caterpillar)	980 G	29 t (5.4 m^3)	460,000,00 €
7	Grader (Graduator)	163 H	16.3 t	285,000,00 €
8	Mixing plant (Eirich-Mischer/Granulierer)	RV15 Vac RV19 Vac RV23 Vac RV29 Vac RV32 Vac	1.200 kg 2.400 kg 4.800 kg 8.400 kg 11.200 kg	131,000,00 € 247,000,00 € 287,000,00 € 529,000,00 € 547,000,00 €

References

[1] Pusch, R., *Waste Disposal in Rock*, Developments in Geotechnical Engineering, Vol. 76, Elsevier: New York, 1994.

[2] Pusch, R., *The Buffer and Backfill Handbook, Part 2: Materials and Techniques*, SKB Technical Report TR-02-12, Swedish Nuclear Fuel and Waste Management Co: Stockholm, 2002.

[3] Grim, R.E., *Clay Mineralogy*, McGraw-Hill: New York, 1967.

[4] Pusch, R. & Kasbohm, J., Can the water content of highly compacted bentonite be increased by applying a high water pressure? SKB Technical Report TR-01-33, Swedish Nuclear Fuel and Waste Management Co: Stockholm, 2001.

[5] Huertas, F. et al., *Effects of Cement on Clay Barrier Performance, ECOCLAY Project*, Final report Contract No. F14W-CT96-0032, European Commission: Brussels, 2000.

[6] Pusch, R., Chemical interaction of clay buffer materials and concrete, Technical Report SFR 82-01, SKB, Stockholm, 1982.

[7] Pusch, R., Zwahr, H., Gerber, R. & Schomburg, J., Interaction of cement and smectitic clay – theory and practice. *Appl. Clay Sci.*, **23**, pp. 203–210, 2003.

CHAPTER 5

Stability analysis of mines

R. Adey & A. Calaon
BEASY and Wessex Institute of Technology, Southampton, UK.

Abstract

This chapter summarizes the modelling work performed to assess the stability of the proposed repositories. The approach adopted is not to consider one specific mine but conceptual mines typical of the type and geological structure found in the EU. Special tools have been developed for the assessment of underground rock stability based on the BEASY software and a complete GiD problem-type programmed for model development and visualization, which allows for a seamless interaction between the two environments. The report summarizes the developments made in the modelling toolkit for the rapid evaluation of potential repository sites and specifically their stability. Two principal rock stability 'cases' have been investigated: one representing a room and a tunnel in a fractured granite matrix 400 m underground, and another representing the room and pillar configuration in a Limestone cave 300 m underground. Results are presented for the displacement, stresses and Mohr–Coulomb criteria, and the different cases discussed.

Conventions

The convention for stresses used in mechanics, BEASY and GiD does not coincide with the conventional one in most geological sciences. In this chapter, work stress is considered positive in case of traction and negative in case of compression. So the maximum and minimum principal stresses described in this work are the minimum and maximum respectively of the typical geological approach (see Mohr–Coulomb criterion). The expressions 'damaged', 'failing', 'not respecting the Mohr–Coulomb criterion' are to be considered equivalent.

5.1 Background and objective of the work

This work uses computational models to predict the performance of the proposed repository. The main objective is to determine the stability and capacity of reference mine repositories to provide safe and permanent isolation of hazardous waste. Stress analysis of the rocks was performed in order to estimate the stability of the repository.

5.1.1 Concept model for prediction

The main objective of this project is to answer the question whether it is possible to use abandoned mines for disposal of hazardous waste with negligible pollution of the environment. Predictive models have been developed to provide quantitative data on the stability and isolation capacity of typical mines in the EU.

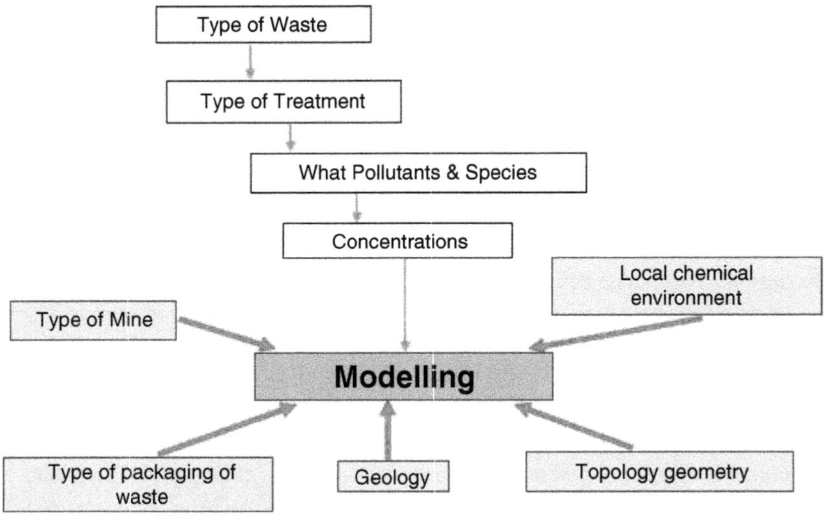

Figure 5.1: Inputs into the numerical modelling of the mine repositories.

Figure 5.2 shows the components of the modelling procedure, where adequate models for the rock stability, flow and transport processes are required as well as other input that will define the containment structure geometry and permeability, geological structure and chemistry of contaminants and environment.

5.1.2 Mine disposal concept

A sketch of the near field geometry is shown in Fig. 5.3. It shows the waste material, a clay barrier and the disturbed rock zone.

The rock contains fracture zones ranging in size between a few hundred meters to a few mm. In the disturbed rock zone a large number of interconnected

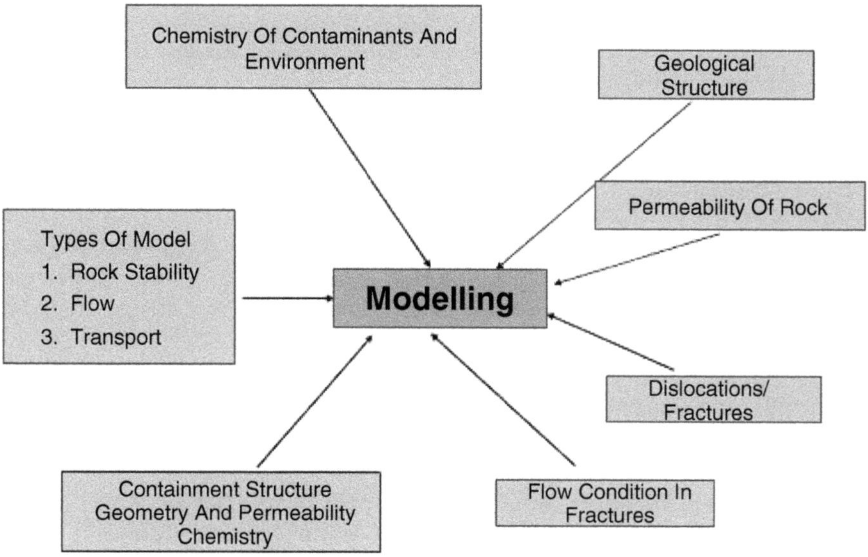

Figure 5.2: Components of the modelling procedure.

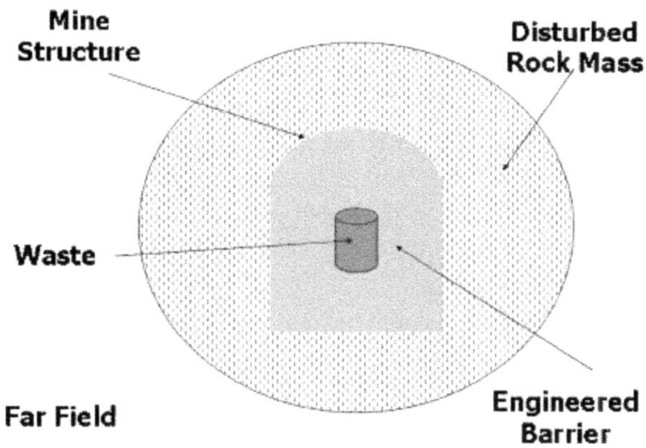

Figure 5.3: A sketch of the geometry of the near field.

fracture zones are present. Beyond this zone the far field starts where only the fracture zones of large scale are of importance. There are different models defined for the near field and for the far field zones.

5.2 Modelling methodology

The stability models were based on the boundary element method (BEM). BEM has a number of advantages for modelling media such as rock, as only the fracture zones are required to be described with elements. The homogeneous media between the fracture zones can be modelled as blocks of material whose boundary is described by the elements on the fracture zones. In the case of Limestone where the fracture zones are not considered so important the elements describe the surfaces of the rooms and the external boundaries. Therefore it needs a much smaller amount of geometrical discretization if compared with analogous finite element models (FEMs).

The presence of different material properties, the geometry of excavation-disturbed zone (EDZ), and the extensive fractured media in the Granite case, make it necessary to have an efficient tool to generate many 'zones' in a single BEM model (Note: zones are regions of the model with different material properties).

A specialized geometry and input file generator was found in the GiD software (http://gid.cimne.upc.es/). Since GiD is a general tool pre- and post-processor and solid modeller, an extensive customization has been necessary to make it able to deal with BEM calculations and with the very special needs of underground modelling, often very different from the ones in mechanics.

Many facilities have been introduced in GiD and a complete problem-type (as the complete set of customizing files is called inside the GiD environment) has been programmed.

One of the major advantages of using a solid modeller as pre-processor is the possibility to have the critical information (in multi-zone BEM) about which volume (zone) a surface belongs to. The access to this knowledge (through Tcl functions reading the GiD database) permitted the efficient production of model files with hundreds of zones which otherwise would have been extremely time consuming and costly. The zone in BEM plays the same role of a single element in FEM but can have any shape and is described by a number of elements. In FEM the material properties (homogeneous media) can be assigned to each and every element, whereas in BEM the zone is the homogeneous domain.

Complicated geometries are also possible with the 3D modelling capabilities of GiD. Although some feature not present in GiD would help significantly in underground modelling (and some of them will probably be included in the next release) all the geometries of the present project have been built with GiD only. Some menus and toolbars have been created in the problem-type to speed the model creation and imposition of boundary conditions.

Some limits remain, especially in post-processing with GiD, so that a complete 'picture' of the model is possible only combining BEASY and GiD.

5.2.1 Mohr–Coulomb criterion

In rock mechanics the most used brittle failure criterion is the Mohr–Coulomb criterion. It establishes a limit for the difference between maximum and minimum principal stresses, depending on the minimum itself, often called 'confining stress'.

Here the equation representing the critical state is used:

$$\sigma_1' = \sigma_1 + K\,\sigma_3'. \tag{5.1}$$

In eqn (5.1) σ_1' is the maximum principal stress (eigenvalue of the stress tensor) and σ_3' is the minimum. If σ_1' is bigger, then the linear function of σ_3' on the right of the criterion is not satisfied. In the equation K represents a multiplication factor for the minimum stress and is equal to:

$$K = \left(\frac{1+\sin\phi}{1-\sin\phi} \right). \tag{5.2}$$

σ_{cm} is the uniaxial compressive strength:

$$\sigma_{\text{cm}} = \frac{2c_0 \cos\phi}{1-\sin\phi}. \tag{5.3}$$

This failure criterion is used in geomechanics as an indicator of probable instability in the short- or midterm, depending on the case. When the criterion is not satisfied the rock is likely to break, lose its mechanical strength and cause material fall into the rooms or tunnels. When the criteria is not satisfied on the walls of a room or tunnel, for instance, the wall would probably be unstable and material would eventually lead to rocks falling into the room unless some action is taken to strengthen the rock.

In order to have the criterion visualized in GiD, an extension of the Tcl post-processing routine was written. The field visualized then has a 0 value everywhere apart from the areas where the failure criterion is not satisfied. There the value of the field changes to 1.

5.2.2 EDZ divided into subzones

Any BEM model is numerically sensitive to the shape ratios (the ratio between the longest and the shortest straight lines crossing the zone) of the zones in the sense that high ratios can cause numerical errors, which could spoil the precision of the whole calculation. Since the EDZs are generally thin layers surrounding the excavated spaces, they represent a typical unfavourable case for the BEM method. In order to avoid numerical problems the EDZ around tunnel and room in the granite cases have been divided into better shaped zones (see Fig. 5.4).

5.2.3 Submodelling

For any numerical model any small feature where the solution is rapidly varying, embedded in a much bigger and 'smooth' part, represents a challenge to be calculated. In fact small features need small elements on them, and even smaller if the variation of the result itself (stress is proportional to the 'derivation' of the displacement calculated with BEM) is of interest (possible local failure). On the other hand a big part of the model does not need a particularly fine mesh since there is very little variation there. So two things occur: big elements are placed very near small ones, and a number of very small distances between nodes appear. Both things lead to numerical inaccuracy and instability.

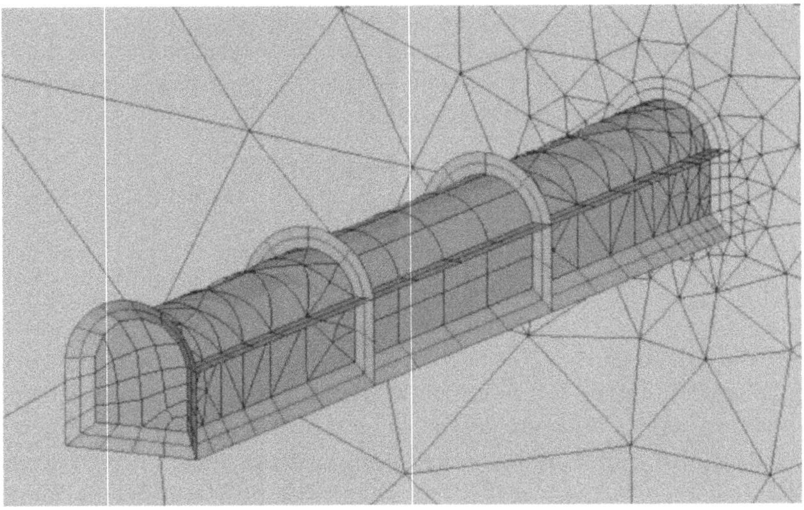

Figure 5.4: EDZ submodelling.

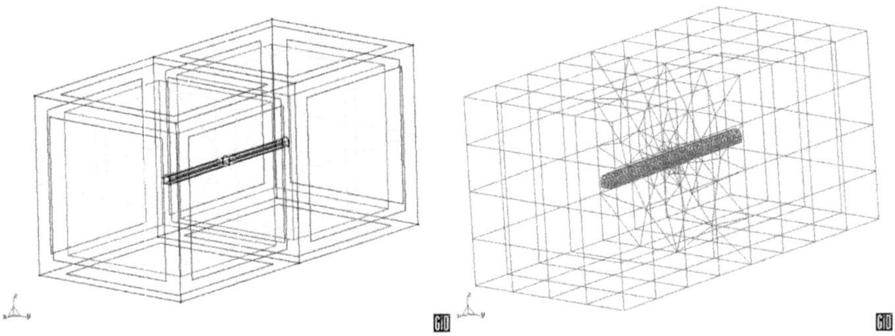

Figure 5.5: Geometry and mesh of the first model without EDZ.

In order to be able to handle the geometry of tunnel and room, each of them crossed by a vertical fracture zone, the EDZ zone is divided in subzones and the big 'cubes' of rock in the granite case, a submodelling technique was used. The new method uses a technique, which can be used not only for this case, but also in the general interaction between FEM and BEM models.

5.2.4 Some experiments to determine the required model details

In the granite case, in order to have an idea of the extent of the critical coulomb zone and of the mesh dimension necessary to represent it with sufficient precision, a number of simplified models were created. These represent a section of the rock with a tunnel in it, split by a fracture zone. Models with and without EDZ (see Fig. 5.5 and Fig. 5.8) were assessed, at the normal tectonic pressure of

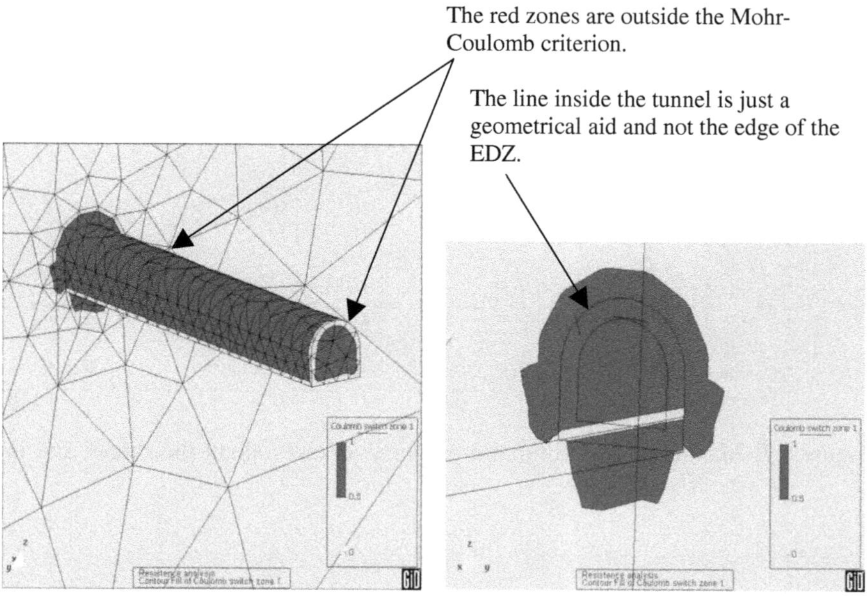

Figure 5.6: Mohr–Coulomb criterion result near the crossing fracture zone in the first model without EDZ.

20 MPa and at an increased one equal to 30 MPa. The tectonic pressure was applied indirectly to the model, by forcing a compression displacement of the side walls. Since the value of the displacements to be used could not be predicted in advance, two cases were considered and a linear interpolation used to obtain the required values.

The first experiment was performed without the EDZ zones and with a tectonic pressure of 20 MPa in all horizontal directions. Some data about the first experiment model:

- degree of elements is: quadratic 9;
- number of points: 40;
- number of lines: 72;
- number of surfaces: 37;
- number of volumes: 2;
- number of quadrilateral elements: 356;
- number of triangle elements: 910.

This experiment showed that for the Mohr–Coulomb criterion all tunnel, walls, ceiling and floor, and its surroundings are unstable (see Fig. 5.6). The extent of the unstable zone inside the rock is about 1.5 m on vault and walls and about 2.5 m under the floor. The maximum and minimum principal stresses (Fig. 5.7) show a tension on the tunnel walls of 3.7 MPa peak. This means that the so-called 'confining stress' cannot reach the walls. A small tension stress appears

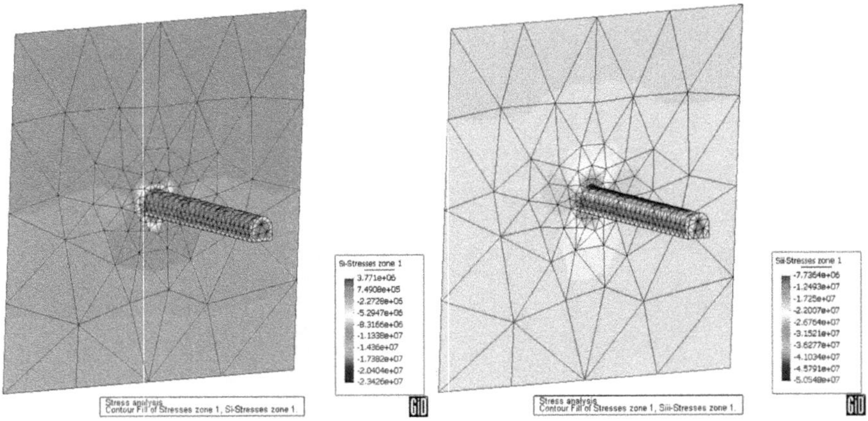

Figure 5.7: Maximum and minimum principal stresses along the tunnel and in the fracture zone.

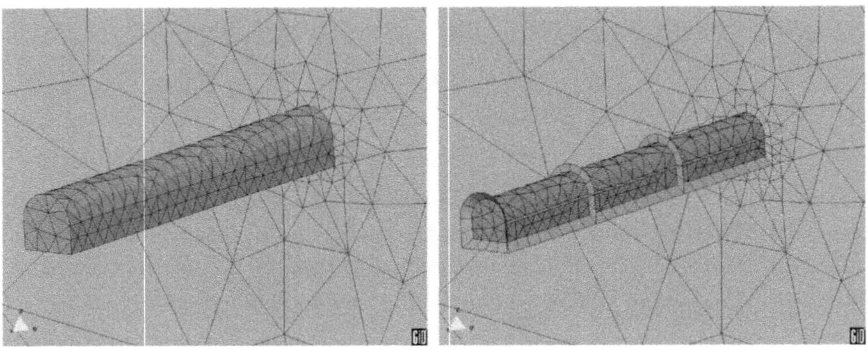

Figure 5.8: Mesh of the second model with EDZ. In both the pictures the 'double' skin of the tunnel necessary to model the presence of the EDZ is visible.

in the rock and endangers the local stability. In the present case not only the sides (where the tension develops) of the tunnel are unstable, but the ceiling and the floor as well (on the floor the maximum principal stress in practically 0 everywhere). This experiment shows also that the necessary mesh resolution around the whole tunnel should be not less than the one used in the example.

A second experiment was performed using the same geometry, but with a 1 m deep EDZ divided into subzones. The dimension of the model in number of degrees of freedom practically doubled, due to the double modelling of the tunnel shape. The new mesh is shown in Fig. 5.8. Some data about this second model:

- degree of elements is: quadratic 9;
- number of points: 68;
- number of lines: 152;

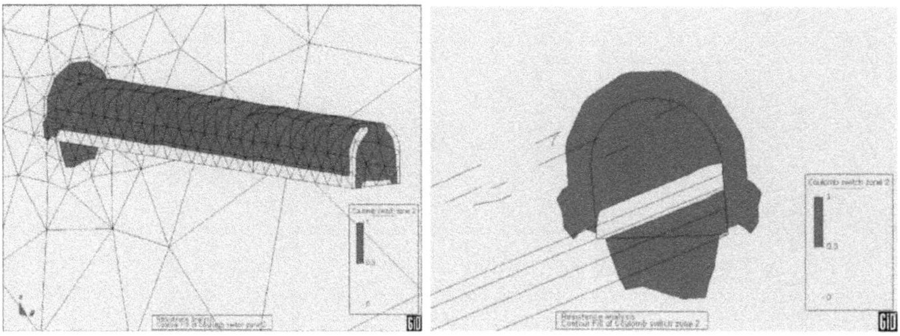

Figure 5.9: Mohr–Coulomb criterion in the model with EDZ.

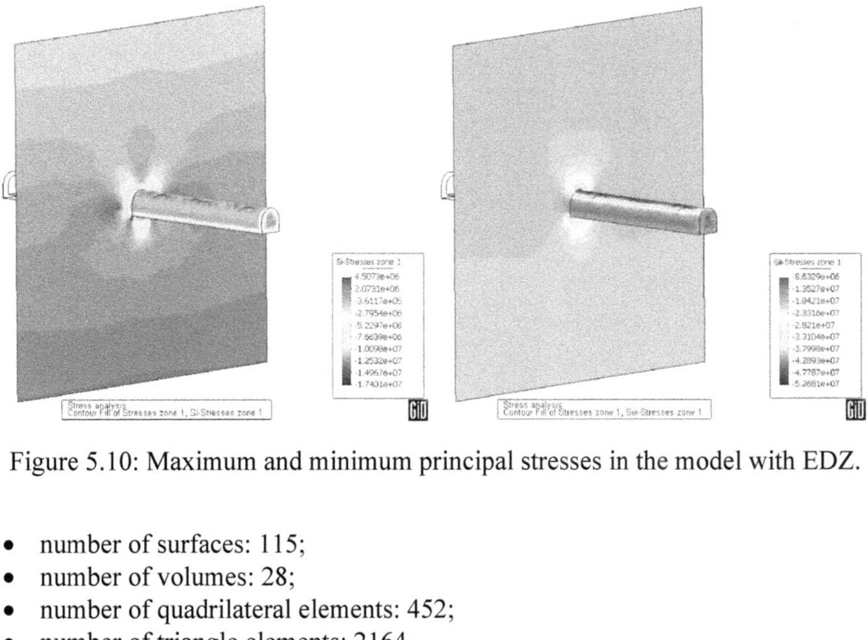

Figure 5.10: Maximum and minimum principal stresses in the model with EDZ.

- number of surfaces: 115;
- number of volumes: 28;
- number of quadrilateral elements: 452;
- number of triangle elements: 2164.

The shape and extension of the failure are practically identical to the ones of the case without EDZ shown in Fig. 5.7. The Mohr–Coulomb result is shown in Fig. 5.9; it is practically the same as in the case without EDZ and the shape of the failure area in the fracture zone is also identical. A comparison can be made between Figs 5.6 and 5.10. The maximum stress, although showing the same pattern, is slightly higher, due to the presence of a minimal support of the EDZ internal layer. The minimum (maximum for geological conventions) is for engineering purposes practically identical in the two cases. This explains also why the extension of the 'Mohr–Coulomb failure zone' is identical for the two cases, and is the reason why we decided that the result of models without EDZ can be taken as

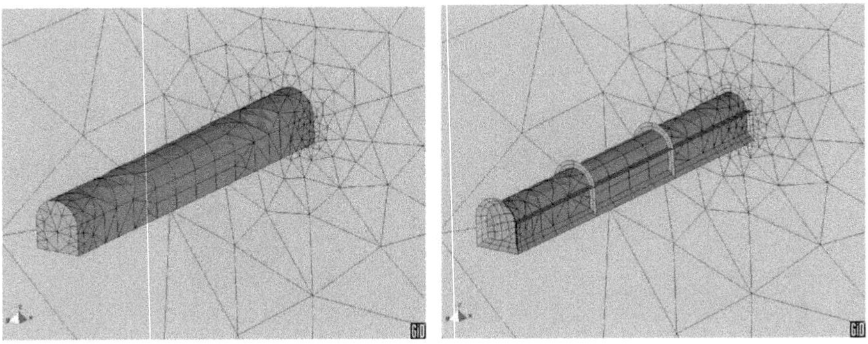

Figure 5.11: The 'reduced' mesh in the third model with external surfaces visible and invisible.

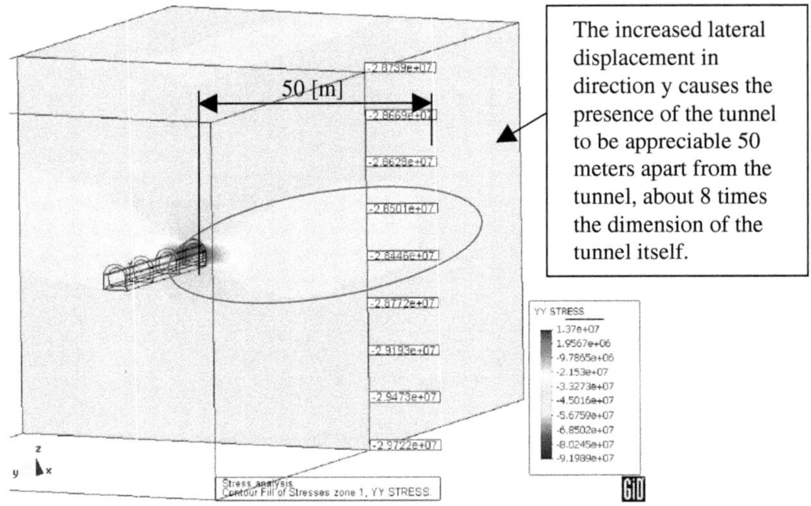

Figure 5.12: *zz* stress. The effect of the tunnel is appreciable at 45 m distance.

valid and representative for the same geometry and boundary conditions but with EDZ. So, in practice, the presence of the EDZ does not change the stability problem.

Another experiment was performed in order to have an idea of the sensitivity of the solution of mesh and tectonic stress distribution. A third model was solved, this time with a coarser mesh (see Fig. 5.11) and the EDZ zone in it (at the moment the model was created we had not still verified the absence of any influence on the stability result of the EDZ). In this model the displacement on the long sides have been increased, resulting than in a higher calculated tectonic stress along the tunnel of slightly less than 30 MPa. The increase was so much that in this case the *yy* stress on the cube side was not linearly increasing with the depth and the influence of the tunnel on the stress distribution was to be filled even 45 m apart, as is recognizable from the Fig. 5.12.

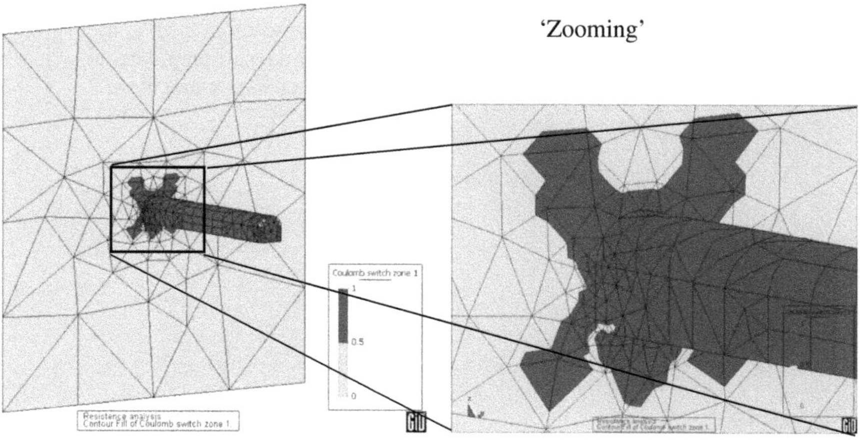

Figure 5.13: Mohr–Coulomb criterion in the third case: the failure zone grows radially.

Some data about the third model is shown below:

- degree of geometrical elements is: quadratic 9;
- number of points: 68;
- number of lines: 152;
- number of surfaces: 115;
- number of volumes: 28;
- number of quadrilateral elements: 971;
- number of triangle elements: 1610.

The Mohr–Coulomb criterion shows in this case failure in a much bigger zone around the tunnel near the fracture zone. It seems that the endangered rock is situated at 45° from the vertical axis as shown in Fig. 5.13. The geometry is probably due to the particular case under study, with symmetrical geometry and boundary conditions.

Figures 5.14 and 5.15 show maximum and minimum principal stresses and a local numerical instability, probably ascribable to the shallow geometry of the EDZ zones around the tunnel.

5.3 Description of cases to be studied and modelling assumptions

5.3.1 Limestone – room and pillar

The geometry of a typical limestone mine is shown in Fig. 5.16. It consists of a regular array of pillars 7 m long and wide, 5 m high repeated over a length of 150 m.

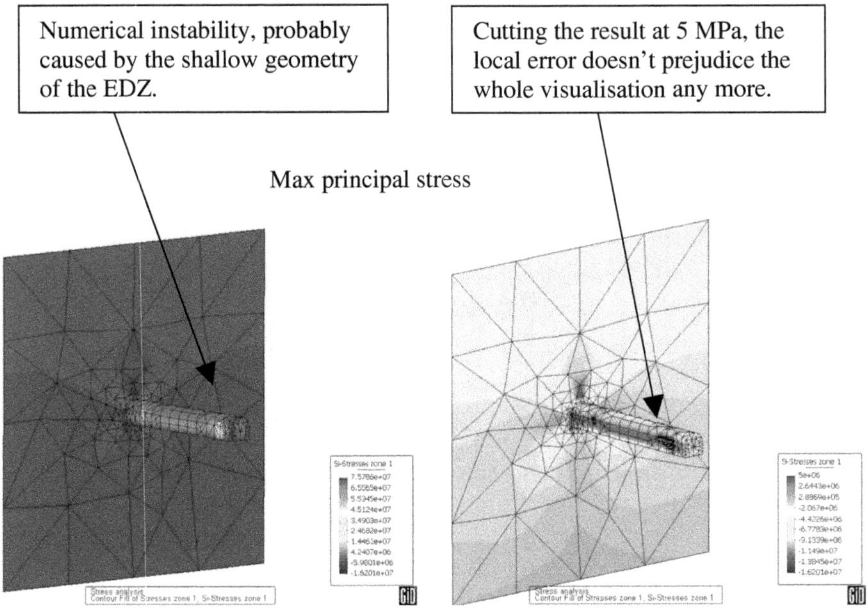

Figure 5.14: Maximum principal stress showing a numerical instability, that does not prejudice the whole result. On the right, the solid fill with the result cut at 5 MPa.

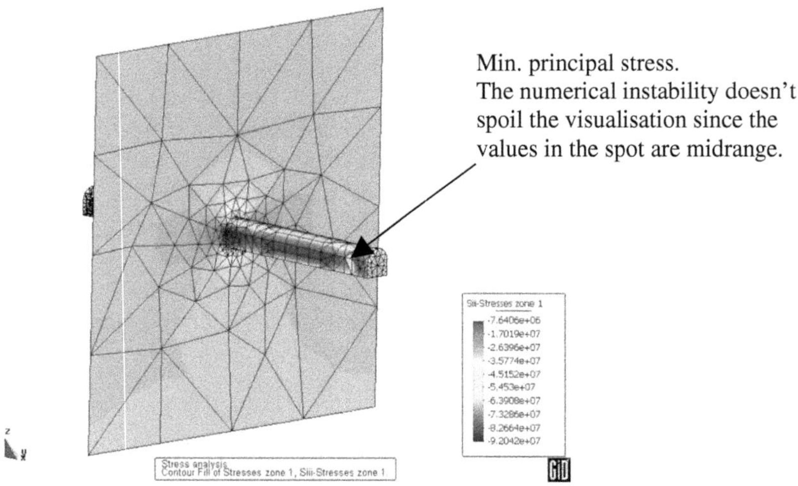

Figure 5.15: Minimum principal stress.

Assuming that the influence of the intact rock around the array boundary on the stress distribution does not extend into the 'central' part of the mine, and considering that no horizontal tectonic forces act, a reduced model of a symmetry cell (room and free space) can be used to represent the behaviour of the

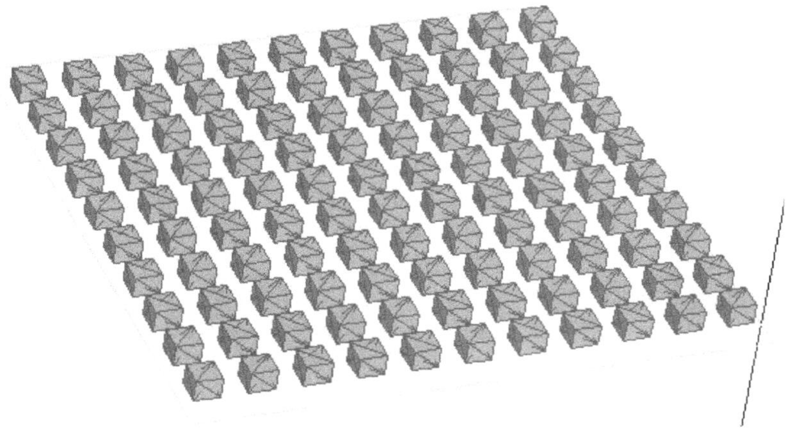

Figure 5.16: Array configuration of a typical limestone room and pillar mine.

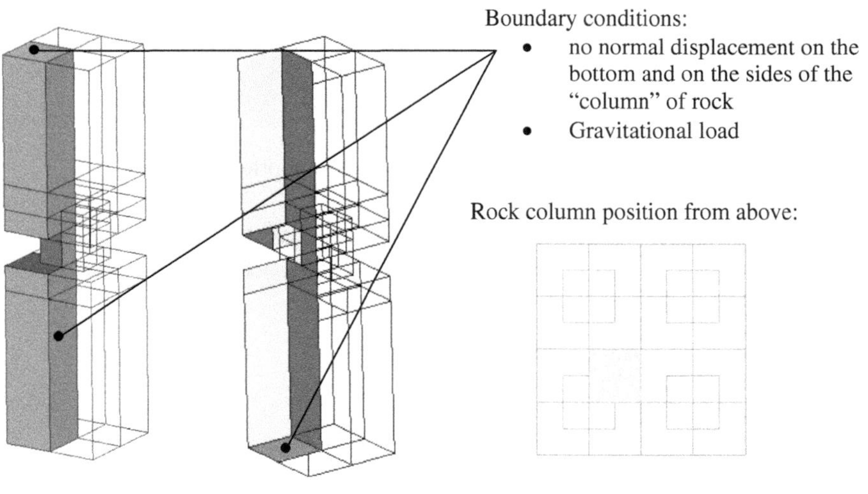

Figure 5.17: Geometry of the columnar model.

excavated rock in the central part of the mine. The reduced model corresponds to a column in an infinite room and pillar array, and its use will be justified by other numerical results in the following.

The geometry obtained is shown in Fig. 5.17. The high of the column of rock above and underneath the mine level was chosen to be 25 m; a distance afterwards confirmed to be adequate by the numerical results showing almost no shear stress on the top and bottom of the column. The force corresponding to the rock mass over the top of the column was applied to the upper surface, the value being 7.15 MPa.

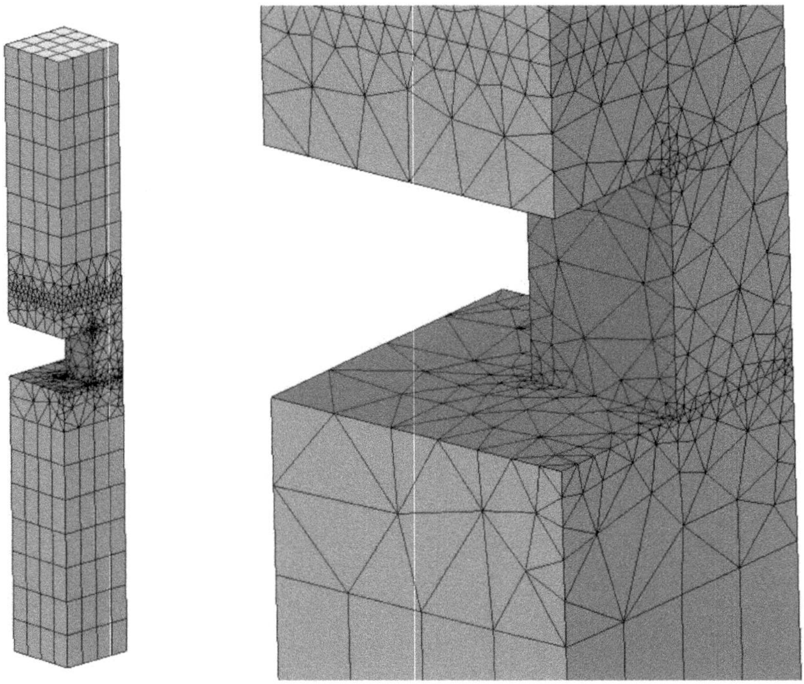

Figure 5.18: Mesh of the columnar model rendered in GiD.

Figure 5.18 shows the mesh applied to this geometry. The mesh was prepared finer near the corners and in the places where stress concentrations occur. Some data about the model and mesh is given below:

- degree of geometrical elements is: quadratic 9;
- number of points: 37;
- number of lines: 67;
- number of surfaces: 36;
- number of volumes: 5;
- number of triangle elements: 2036;
- number of quadrilateral elements: 256.

To investigate the rock stability at the end of the room and pillar array (excavating front), a second model was prepared and solved. The geometry is shown in Fig. 5.19. It corresponds to a section partly inside the mined space, and partly inside the intact rock. But how far should the slice extend to be able to use normal displacement boundary at its end?

In this case it is necessary to extend the model into the undisturbed rock a sufficient distance to achieve the stress conditions for the bulk rock. The dimension chosen was 25 m, as in the previous case, corresponding to 5 times the room high and 3.8 times the pillar spacing. The same distance was used for the high of the column of rock above and underneath the mine.

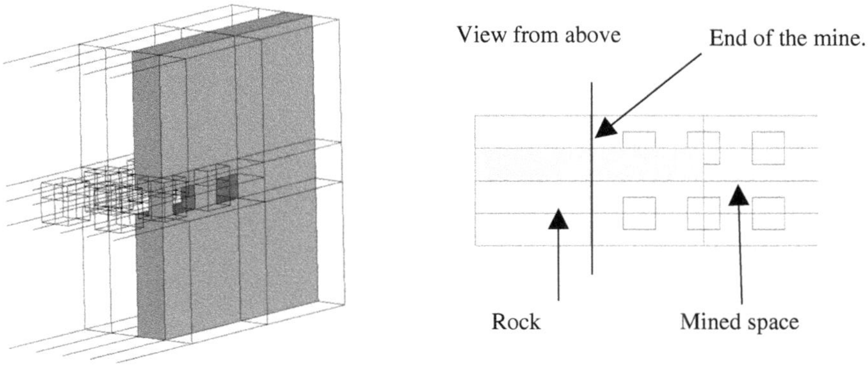

Figure 5.19: The second model representing the end of the mine in a 3D view and a vertical projection.

For the limestone case the EDZ was considered extending vertically above the mine ceiling (not above the columns) for 3 m depth. Some data about the model:

- degree of elements is: quadratic 9;
- number of points: 215;
- number of lines: 426;
- number of surfaces: 54;
- number of volumes: 10;
- number of triangle elements: 5206;
- number of quadrilateral elements: 135.

No fracture zones appear in the rock.

Figure 5.20 shows an overview of the mesh adopted for this geometry. As in the previous case the mesh was finer at corners and near stress concentrations in order to obtain better local accuracy.

5.3.2 Crystalline rock

The geometry of the crystalline (or granite) case is as follows.

The rock mass surrounding room and tunnel was modelled extended to a depth of 800 m below the ground level. A square area of 600 by 600 m represents a horizontal cross section of a parallelogram enclosing the whole model.

The rock mass is cut by three orthonormal fracture zone families, two vertical and one horizontal. For all of them the distance between the discontinuity planes is 100 m, so the rock matrix around room and tunnel is made up by a 3D array of cubes with a 100 m long edge.

As shown in Fig. 5.21, a room and a tunnel, both empty, are located inside three blocks in the 'middle' of the model, with the room ceiling depth at 425 m. Both room and tunnel are crossed by only one vertical fracture zone. The room dimensions are: cross section – 50 (m) × 50 (m); length – 100 (m). The tunnel dimensions are: cross section – 5 (m) × 5 (m); length – 150 (m).

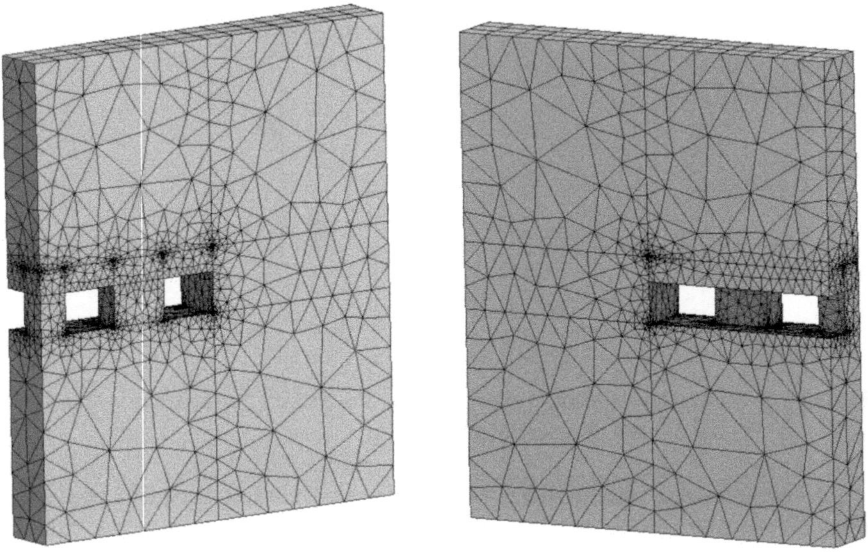

Figure 5.20: Mesh of the side model, seen from the two different sides.

The tunnel is parallel to the length of the room and departs from it starting from the centre of a bottom edge of the room.

Figure 5.22 shows both room and tunnel surrounded in all direction by the EDZ, 3 m deep around the room and 1 m around the tunnel.

Without submodelling the number of degrees of freedom in the model would be 250,000 and too large to be solved in a reasonable time. This conclusion still applied when attempts were made to limit the mesh refinement and using techniques like the transition mesh for passing from a coarse to a refined one. Therefore the submodelling strategy was chosen, and two different models prepared from the large model.

A worst-case situation was considered in that the friction in the fracture zones was assumed to be extremely small and thus neglected.

5.3.2.1 Global (outer) model

An 'outer' model was created starting from the complete one where only the 'rock cubes' where represented. The model consists of 288 volumes and is shown in Fig. 5.23.

The bottom surface has been constrained vertically, and the sides moved inward, squeezing the rock blocks. To have an equivalent stress of about 20 MPa on the sides, a normal displacement of few centimetres was applied to the eternal 'walls' of the model. Due to the large number of blocks present in the model, in comparison with the previous ones, instability occurred, and a number of models showed a singular matrix. The cause was the lack of vertical restraint despite the gravitational load.

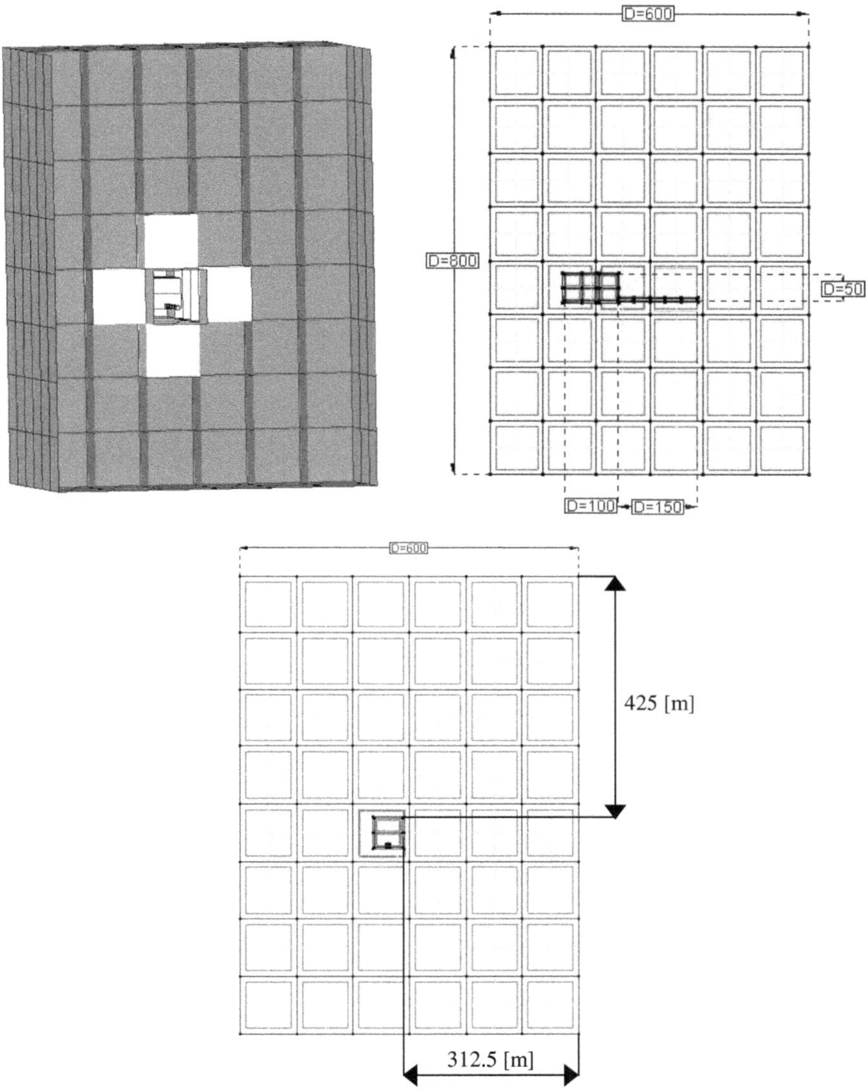

Figure 5.21: Scheme of the complete model for granite.

It was necessary to apply a small vertical restraint on the upper surface [spring modulus ($k = 10^6$ Pa)] which eliminated the problem. The initial choice of using force boundary condition on two walls only and squeezing the model on the opposite side was not used in the present case, to avoid asymmetries on the submodel. Obviously, with a symmetric boundary condition set, as the one used, the opening of pipes in the lines where two fracture zone meet would not be seen.

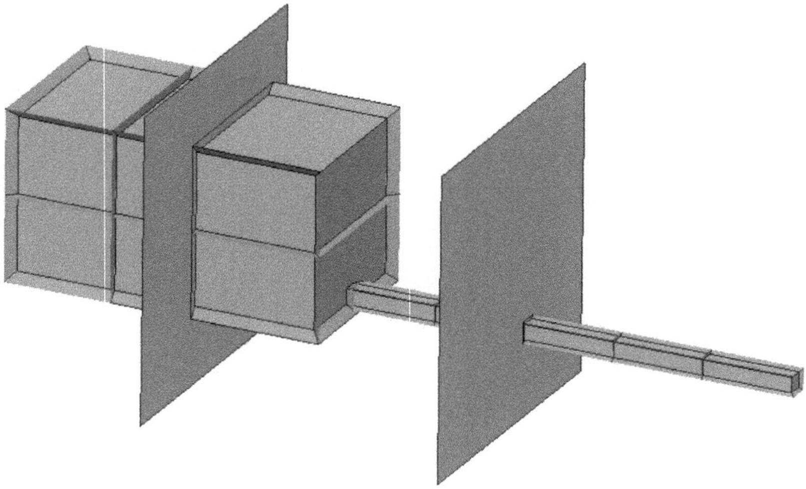

Figure 5.22: Room surrounded by the multi-zone EDZ and the two vertical fracture zones.

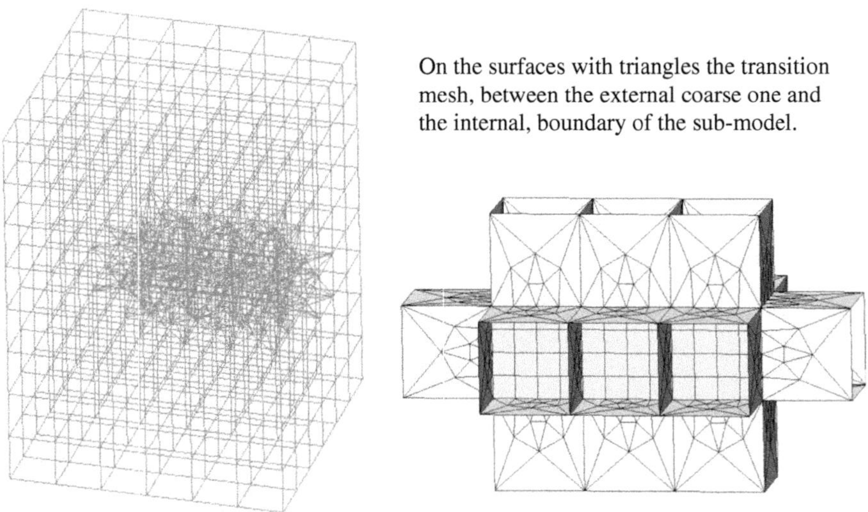

Figure 5.23: Outer model with refined mesh in the part surrounding the submodel.

Some data about the mesh is given below:

- degree of geometrical elements is: quadratic 9;
- number of points: 441;
- number of lines: 1148;
- number of surfaces: 996;

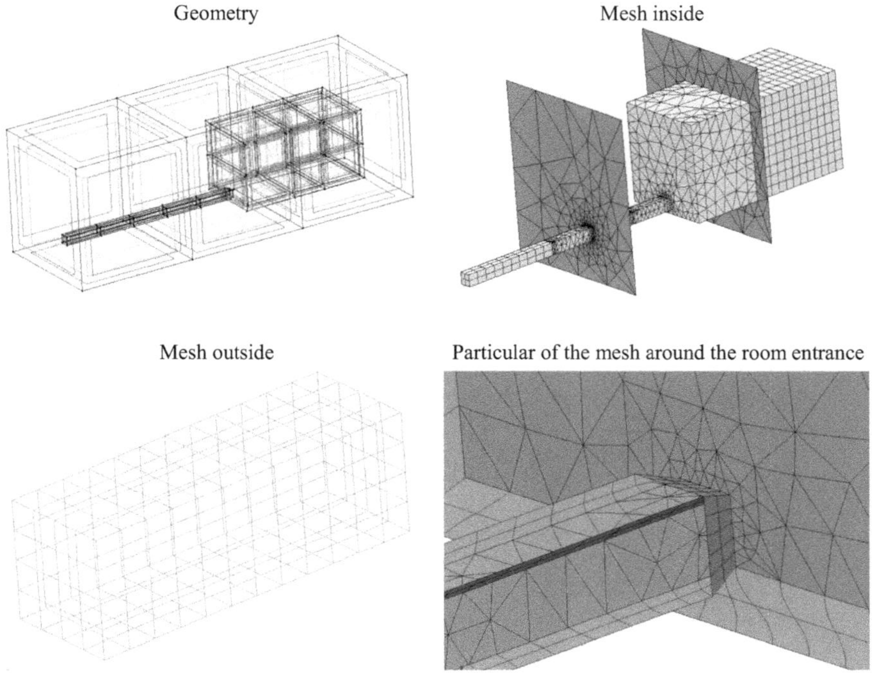

Figure 5.24: Submodel geometry and mesh.

- number of volumes: 288;
- number of quadrilateral elements: 1188;
- number of triangle elements: 720.

5.3.2.2 Submodel

The submodel, sharing its external surface with the 'outer' one, is depicted in Fig. 5.24.

Due to the presence of the EDZ, completely surrounding the inner spaces, and the kind of mesh necessary to 'see' the possible failure near the fracture zones, the number of degrees of freedom in this model is very high for a BEM calculation: 252,000.

Some data about the mesh:

- degree of elements is: quadratic 9;
- number of points: 112;
- number of lines: 253;
- number of surfaces: 190;
- number of volumes: 46;
- number of quadrilateral elements: 2408;
- number of triangle elements: 3140.

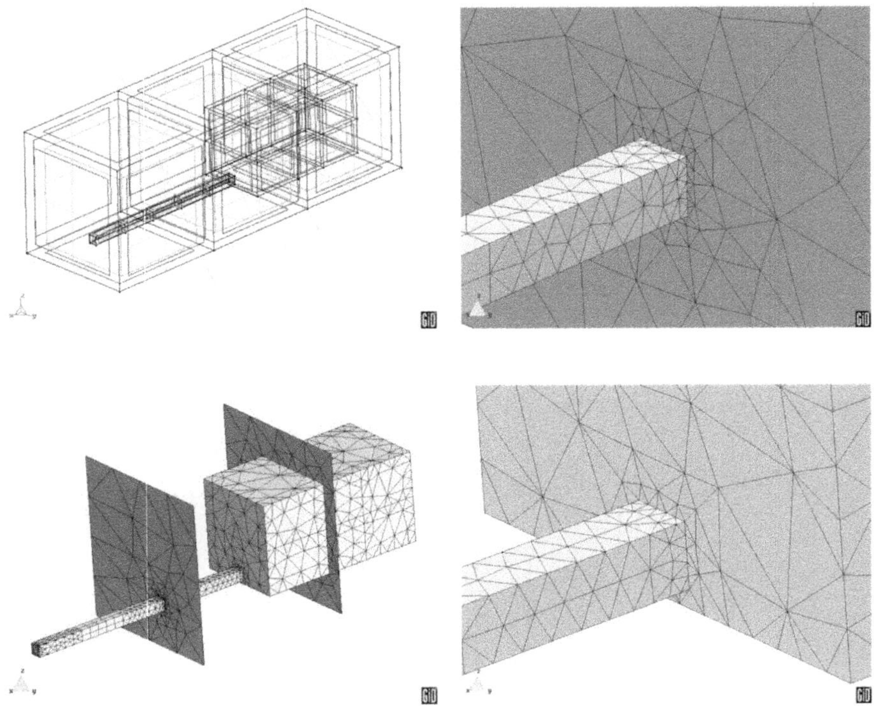

Figure 5.25: Submodel without EDZ.

Since the experiments described previously showed that the presence of the EDZ would not significantly change the stability conditions, a simplified model was prepared (see Fig. 5.25), in which no EDZ was modelled, assuming so a worst case.

The model data is shown below:

- degree of elements is: quadratic 9;
- number of points: 64;
- number of lines: 116;
- number of surfaces: 59;
- number of volumes: 3;
- number of triangle elements: 1902;
- number of quadrilateral elements: 264.

As explained before, the displacement boundary conditions have been calculated from the 'big' model and introduced in the data file.

This model, much smaller that the preceding one, was solved successfully.

5.4 Material properties

In Table 5.1 the material properties to be used in the calculations are given.

Table 5.1: Material properties used.

Rock parameters	Granite	Limestone
Intact rock		
E	50 GPa	12 GPa
ν	0.3	0.25
c_0, φ	$c_0 = 1$ MPa $\varphi_0 = 35°$	$c_0 = 0.5$ MPa $\varphi_0 = 35°$
EDZ		
E EDZ	5 GPa	1.2 GPa
ν EDZ	0.3	0.25
c_0, φ in EDZ	$C_0 = 0.1$ MPa $\varphi_0 = 30°$	$C = 0.05$ MPa $\varphi_0 = 30°$

5.5 Results of stability analysis

5.5.1 Case of mine in limestone

5.5.1.1 Case 1: Centre of the mine

The vertical z displacement near the pillar is pictured in Figs 5.26 and 5.27. As can be seen, the ceiling of the room tends to sink far from the support of the pillar and the pillar itself sinks a little inside the matrix of the limestone.

Comparing Fig. 5.26 with Fig. 5.27 it is possible to see that the sinking of the ceiling is rapid near the pillar and reaches a maximum in the point between the pillars (Fig. 5.26) and in the centre of the 'squares' between four pillars (Fig. 5.27). The displacement value is similar in the two cases and is about 12 mm, slightly more in the second position.

No changes in stress distribution appear near the two ends of the column, as visible from Fig. 5.28. This justifies the initial assumption of modelling only a limited column of material.

The results for the Mohr–Coulomb criterion on the column of material in the middle of the array are shown in Figs 5.29–5.31.

The whole EDZ appears to be damaged and the Mohr–Coulomb criterion is not satisfied underneath the pillar in a 'reversed dome shaped' volume. The room floor is damaged to a depth of about 1 m and the pillar appears to be safe in its volume (Fig. 5.29), but fractured on the surface, where a light tension could easily break the rock.

The upper part of the column, above the ceiling shows a similar pattern to the one underneath the pillar: the extension of the failing zone is about 8 m above the end of the EDZ inside the intact rock, whereas above the excavated space the rock is intact and only the EDZ fails.

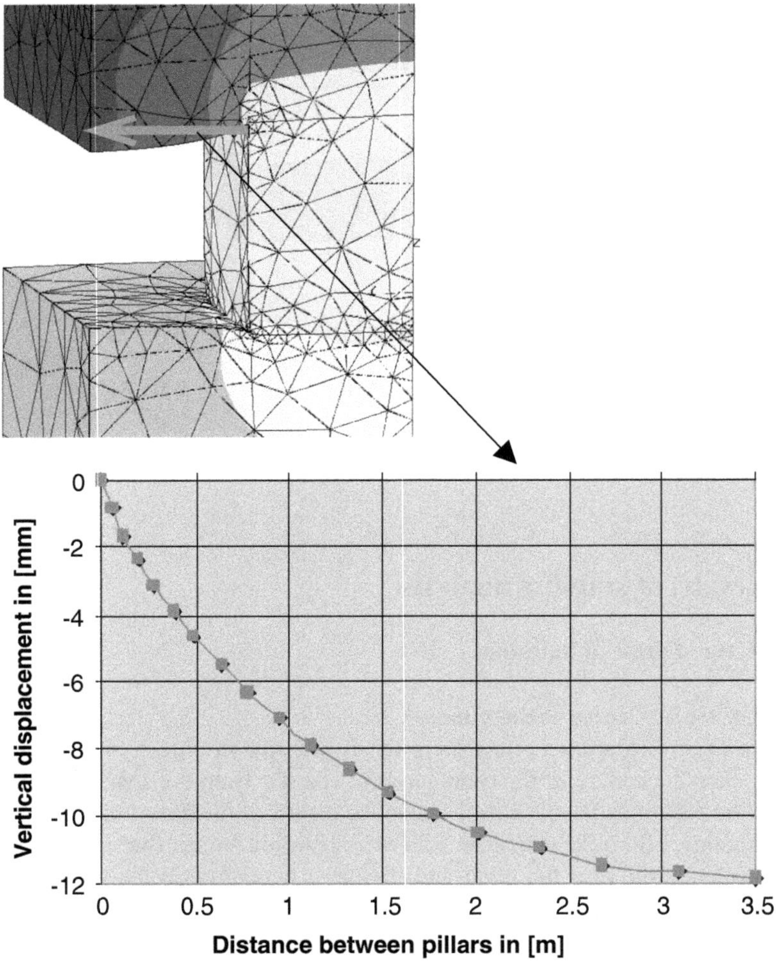

Figure 5.26: *z* displacement. Solid fill on a deformed mesh (top) and graphic (bottom).

Figure 5.32 shows the distribution of principal stresses inside the column (half section). The room space is on the right of the pictures.

The most critical zone in the pillar appears to be near the surface of the pillar at about 2/3 of its high. In case of insufficient support, the pillar would probably start to lose its 'skin' from the superior part.

In Figure 5.33 the distribution of the principal stress in the EDZ is shown. It appears clear that far from the column the contribution of the EDZ to the support of the column of rock over it is minimal. The principal becomes elevated in the intact rock column 6 or 7 m above the end of the EDZ.

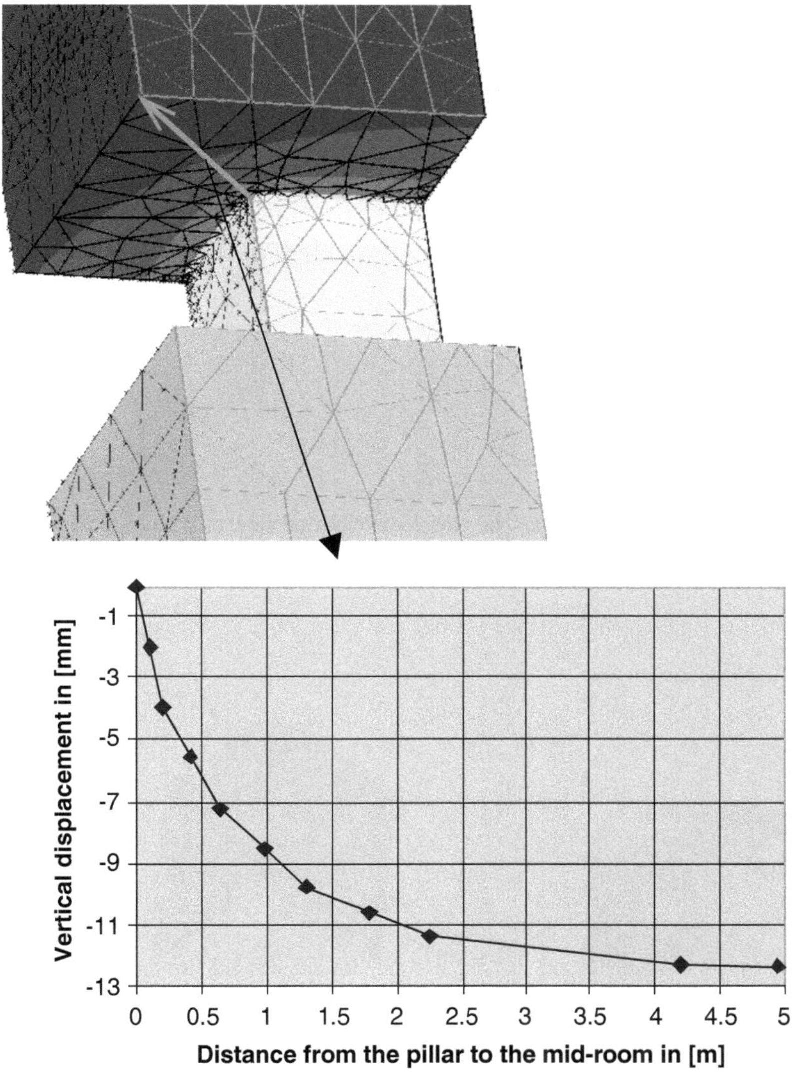

Figure 5.27: z displacement. Solid fill on the deformed mesh (top) and graphic (bottom).

5.5.1.2 Case 2: On the edge of the mine

On the edge of the mine the situation is similar to the one in the middle of the room, where the extension of the fractured zone (where the Mohr–Coulomb criterion is not fulfilled) is about 3 m above the ceiling and just a thin layer in the floor of the rooms. Due to the support of an intact matrix of rock at the end of the mine, the extension of the fractured zone on the ceiling doesn't reach the front wall of

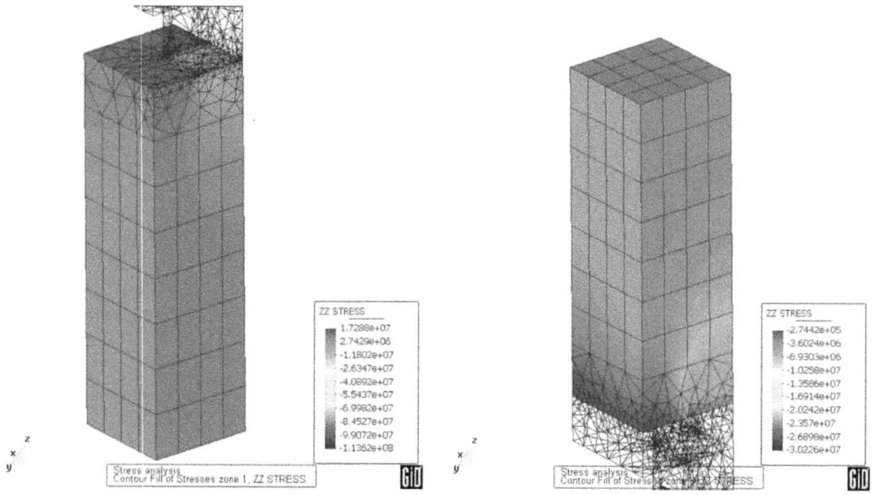

Figure 5.28: Sigma *zz* stress. The stress distribution does not change near the top and bottom of the column.

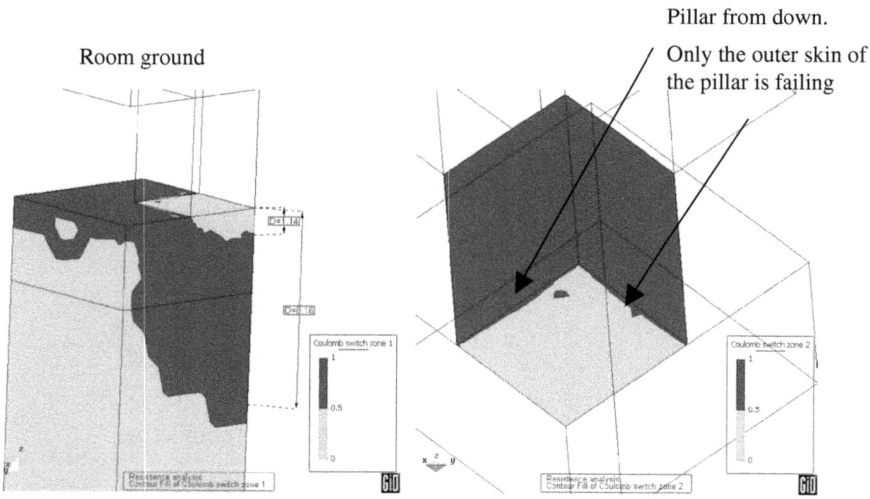

Figure 5.29: Mohr–Coulomb criterion underneath the pillar and in it. The failure zone extends deeply into the rock for about 8 m.

the mine. The assumption about the zone of interest for a realistic application of symmetry boundary conditions is confirmed as adequate by the distribution of the stresses that do not change as far as 25 m inside the rock matrix and shows symmetry already for the first pillar near the end of the mine (see Fig. 5.34).

From Fig. 5.34 is possible to appreciate the gradual decrease in the sinking of the ceiling, as nearer the free space gets to the end of the mine. The maximum

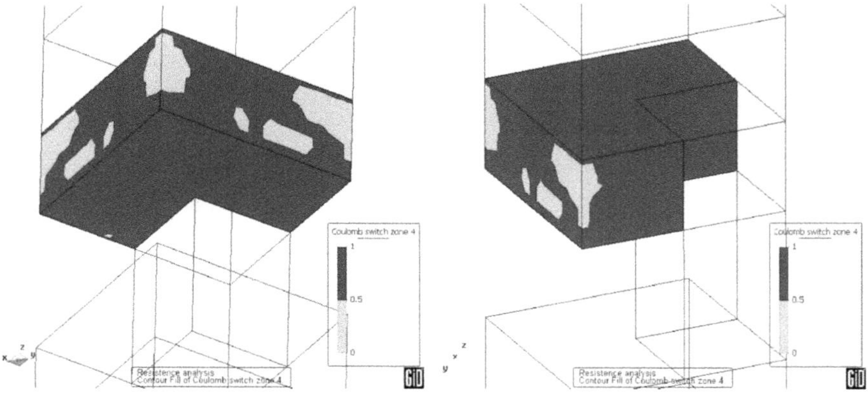

Figure 5.30: Mohr–Coulomb criterion in the EDZ. This zone is completely 'fractured'.

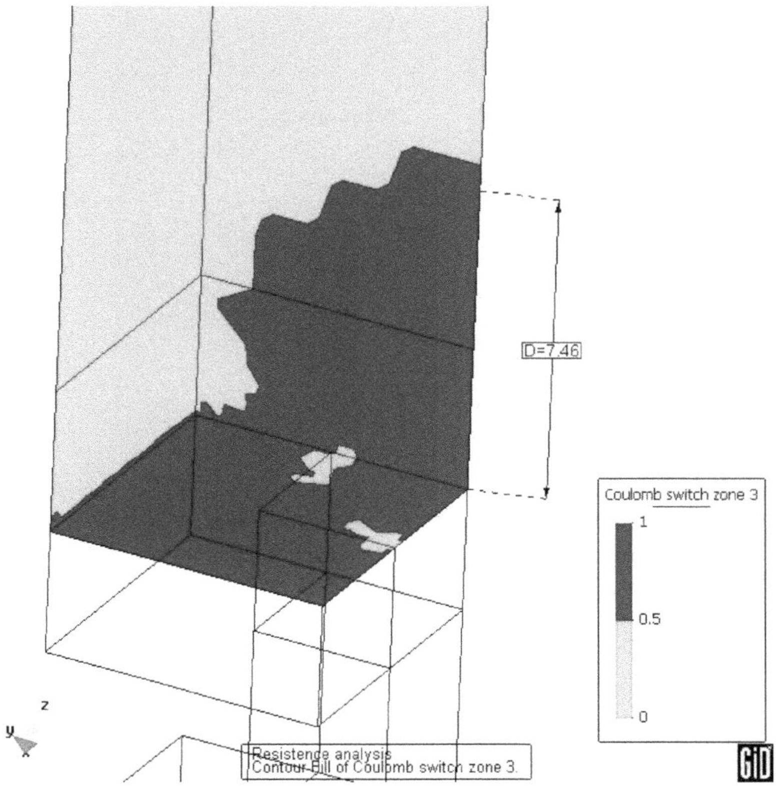

Figure 5.31: Mohr–Coulomb criterion in the upper part of the column, above the ceiling.

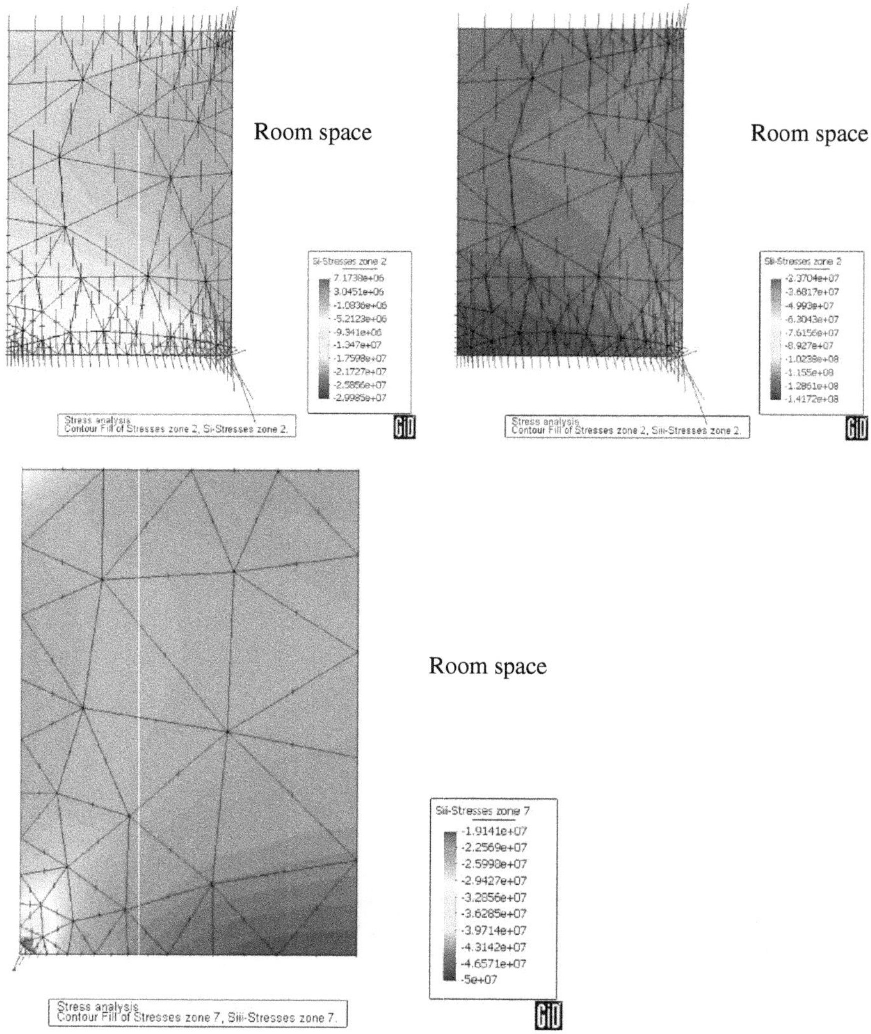

Figure 5.32: Principal stress distribution inside the pillar. The arrows show the principal directions. The last picture shows the minimum stress but rescaled to void the disappearing of the 'blue colour' near the concentration at the bottom.

sinking (relative to the corner far from the mine end) in the first free space near the end wall is about 9 mm and the difference between the two corners about 3 mm. In the second free space the sinking is about 10 mm and the curve is almost symmetrical, showing a corner difference of about 1 mm. The trend in the behaviour of the solution is such that the difference in the next free space would be practically negligible, justifying the assumption of symmetry.

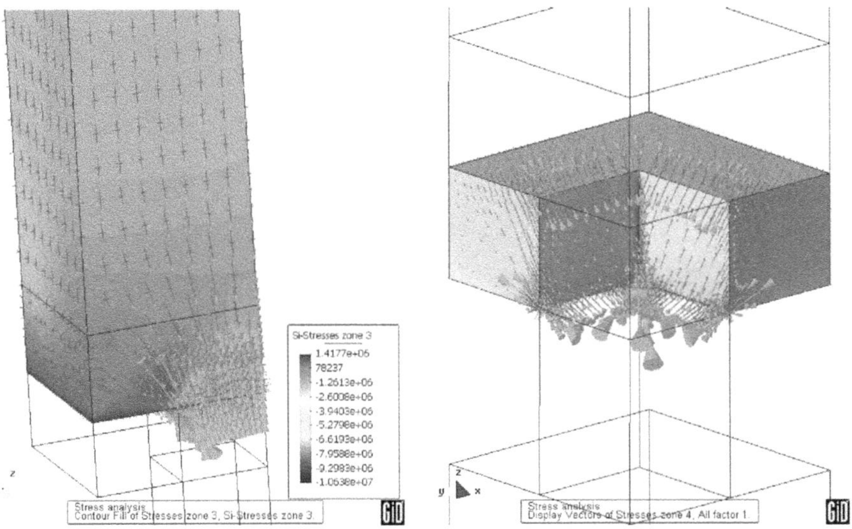

Figure 5.33: Left: Principal stress in the upper part of the model, together with vectorial representation of the 3 principal directions and stresses. Right: Principal stress in the EDZ. A concentration appears in the internal edge.

From Figs 5.35–5.38, showing the Mohr–Coulomb results, it appears that the whole volume of the EDZ is damaged, but the extension of the damage does not go farther than 2.8 m inside the intact rock above.

Figures 5.40 and 5.41 clearly show how locally the principal stresses change direction and intensity near the pillars and the weaker EDZ, in order to transmit all the weight of the above rock to the pillars only.

As in the previous case of the column, the stress concentrates on the pillars starting from about 6–7 m above the ceiling. The Mohr–Coulomb failure extends all over the EDZ and in a thin layer above, making bolting or using shotcrete necessary procedures. Figures 5.39–5.44 show the stress results obtained around the model.

5.5.2 Crystalline rock

5.5.2.1 Case 3: Global model

The results of the global model are very simple to interpret, since the whole geometry shrinks horizontally, and, due to the confining pressure, it raises up slightly in the upper part (geological movement). The springs on the upper surface avoiding the instability produce a tiny compressive force downward (like the gravity) for the stress distribution in the model.

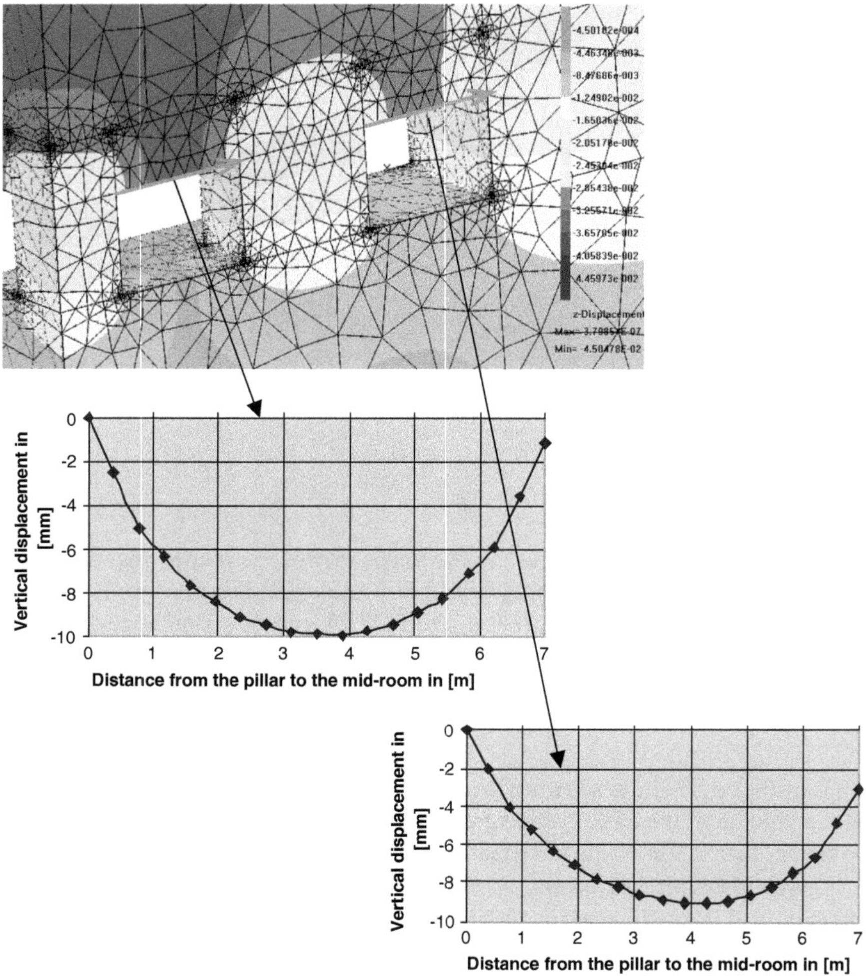

Figure 5.34: z displacement and relative graphs on the pillar side. The two graphs take as a 0 for the z displacement the corner on the left (far from the end of the mine).

In order to obtain the tectonic 'pressure' of approximately 20 MPa at the mid-room depth more than one calculation was performed. With linear interpolation between the two results, the required tectonic pressure was obtained. The solution is shown in Figs 5.45 and 5.46.

A lateral compressive stress linearly increasing with the depth results form the calculation, and its intensity varies form about 14.3 MPa near the surface, to about 23.4 MPa at the bottom.

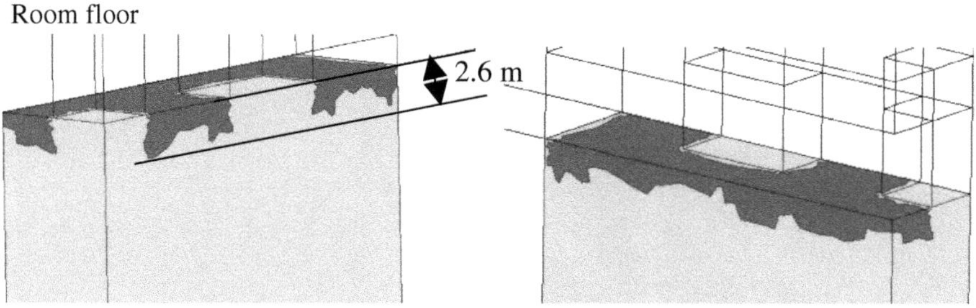

Figure 5.35: Mohr–Coulomb result in the zone underneath the room. The depth of the fractured zone is about 2.6 m.

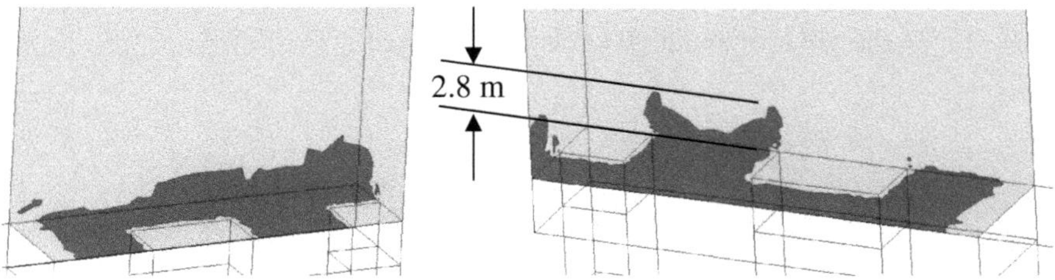

Figure 5.36: Mohr–Coulomb result over the EDZ. The failure penetrates the EDZ above for a maximum of 2.8 m.

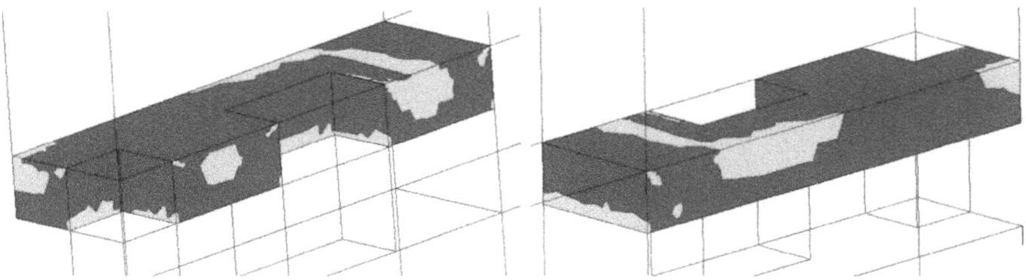

Figure 5.37: Mohr–Coulomb criterion in the EDZ. This zone appears extensively damaged in the EDZ.

5.5.2.2 Submodel

The displacement of the outer model on the surface shared with the submodel was applied as boundary conditions to the submodel without EDZ, assuming the presence of room and tunnel does not have any influence on the outer model solution.

The mesh points of outer and submodel coincided in this case but the software passing BEASY results (sf_ files) to a .dat BEASY input file is now capable

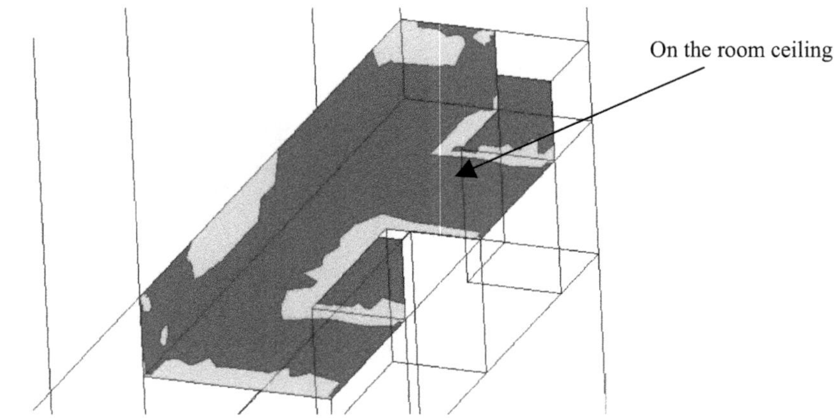

Figure 5.38: Failure in the EDZ as seen from the room. Only a thin layer around the pillar of about 80 cm is not damaged.

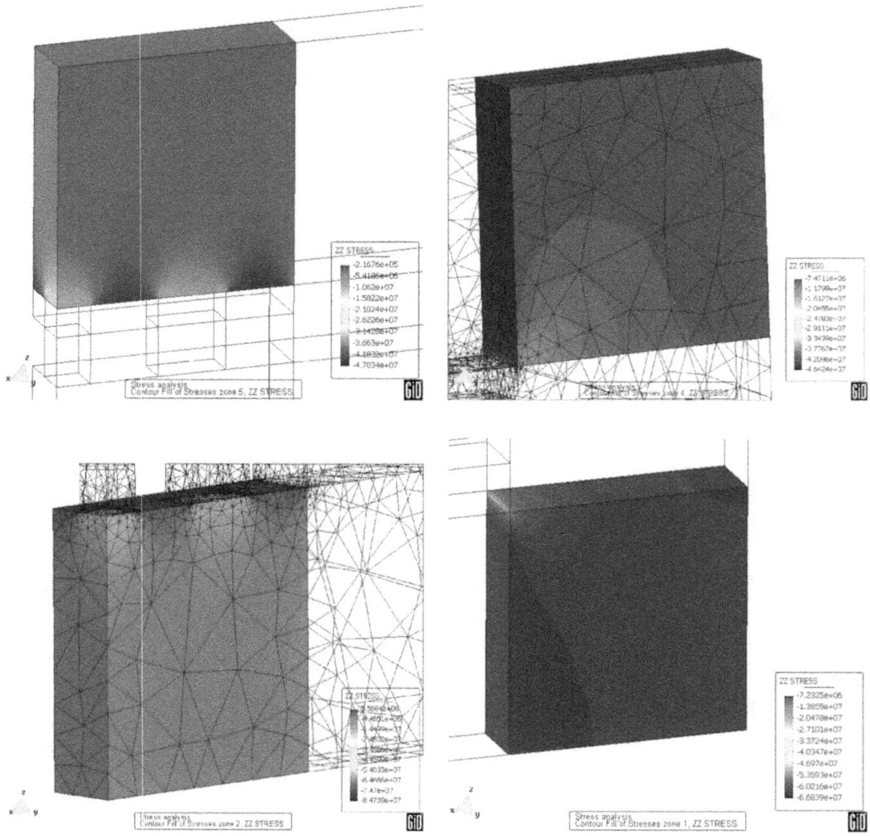

Figure 5.39: Sigma *zz* around the mine. The distribution varies only in the centre and the model appears to be extended enough for the boundary conditions used.

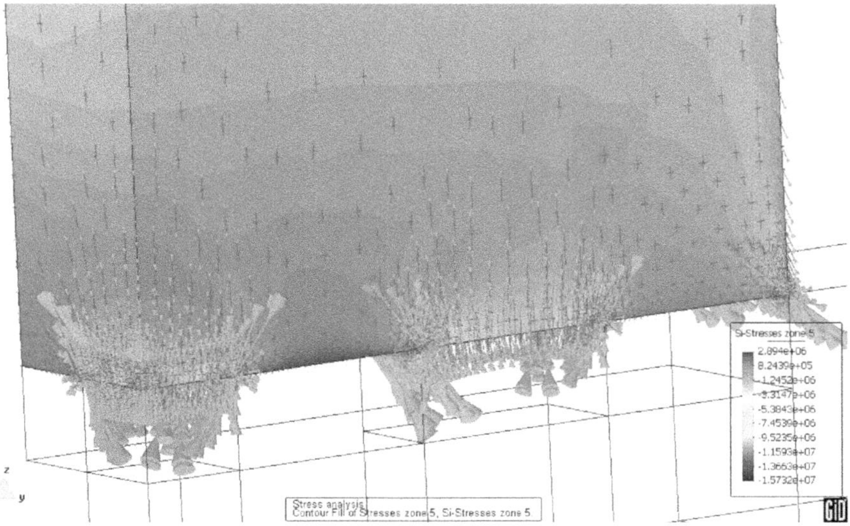

Figure 5.40: Principal stress concentrating on the pillars of intact rock above the EDZ.

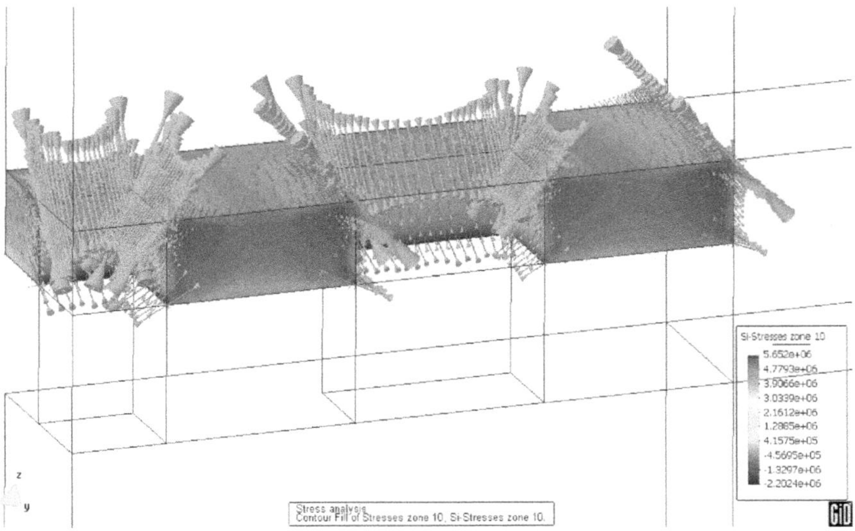

Figure 5.41: Principal stress in the EDZ. In the re-entering corners appears a traction stress (red colour of the scale on the right) but not visible in the picture.

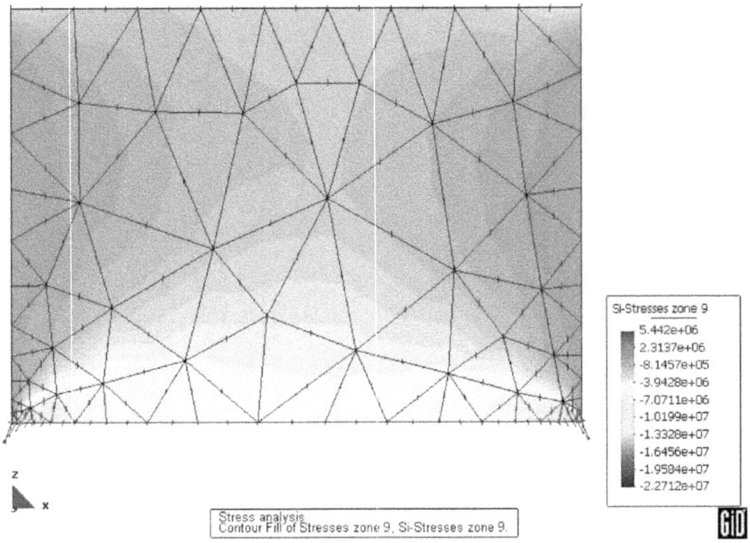

Figure 5.42: Maximum stress in the first pillar near the end wall. The small line represents the direction of the principal stresses, practically everywhere directed vertically, apart from the two low corners, where a stress concentration occurs. The asymmetry of the result is due to the asymmetry of the model itself.

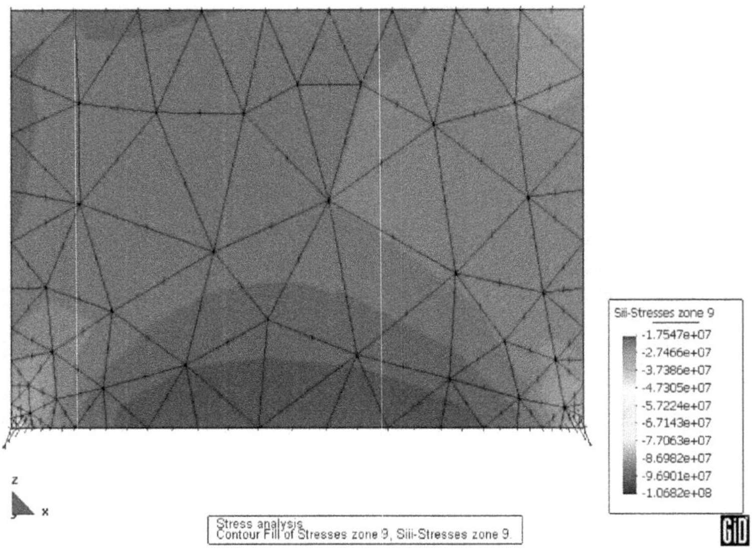

Figure 5.43: Minimum principal stress in the pillar near the end wall (unfortunately the stress concentration near the two lower corners takes 'all the blue').

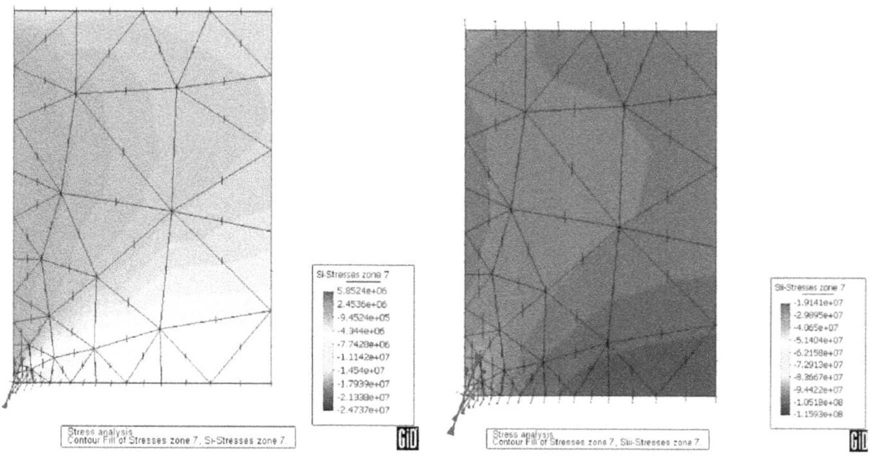

Figure 5.44: Maximum and minimum stresses in the half pillar section far from the end wall. The situation is very similar to the one in the columnar solution.

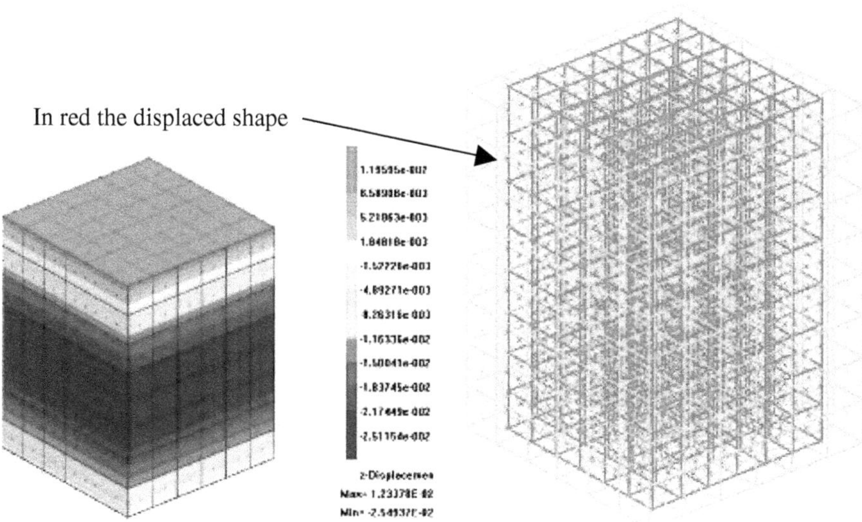

Figure 5.45: Outer model displacement results.

of interpolating and so mapping the results on one mesh to input data on a different one.

The model was solved and the stress results are presented in Figs 5.47–5.51.

In Fig. 5.47 the principal stresses on the tunnel far from the room are shown. As previously seen in the experimental model, a light tension (maximum principal stress) develops along the walls of the tunnel and the minimum

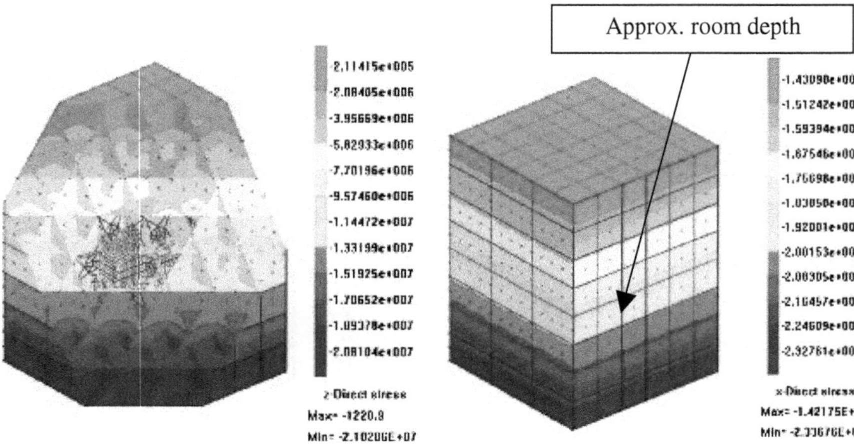

Figure 5.46: Sigma *zz* and *xx* inside the model. The internal mesh transition for submodelling is visible in the picture on the left.

principal stress is maximal there, so limiting the effect of the tension for the Mohr–Coulomb failure. The shape of the stresses near the fracture zone was similar to the results already presented for the 30 MPa tectonic stress along the tunnel (Fig. 5.13).

The maximum principal stress in the 'face wall' of the room near the tunnel shows some tension around the room walls, but its value is very low (of the order of 0.2 MPa).

The maximum principal stress is particularly high on the side walls of the tunnel (Fig. 5.47, maximum value 3.7 MPa tension and Fig. 5.29, maximum value 6.5 MPa). The kind of distribution of the stress on the tunnel is very similar to the one obtained in the experiment with 20 MPa tectonic stress.

The Mohr–Coulomb results are shown in Figs 5.50 and 5.51. The wall of the tunnel appears to be damaged almost on its entire length, and the most critical part is the one near the room, where the extension of the failure goes from ceiling to the floor.

Around the room the most dangerous area appears to be the entrance wall, in correspondence with the tunnel access. The upper part of the wall shows failure, and would probably be the first to cause dangerous rock fall. The rest of the room appears to be safe, but the calculation with an increased tectonic stress (Figs 5.52–5.56) shows it completely collapsing. The results suggest that the state calculated with 20 MPa tectonic stress is very near the load at which the instability propagates form the front wall to the rest of the room walls and ceiling.

The same model was solved as well with an increased tectonic stress corresponding to about 30 MPa (Figs 5.52 and 5.53).

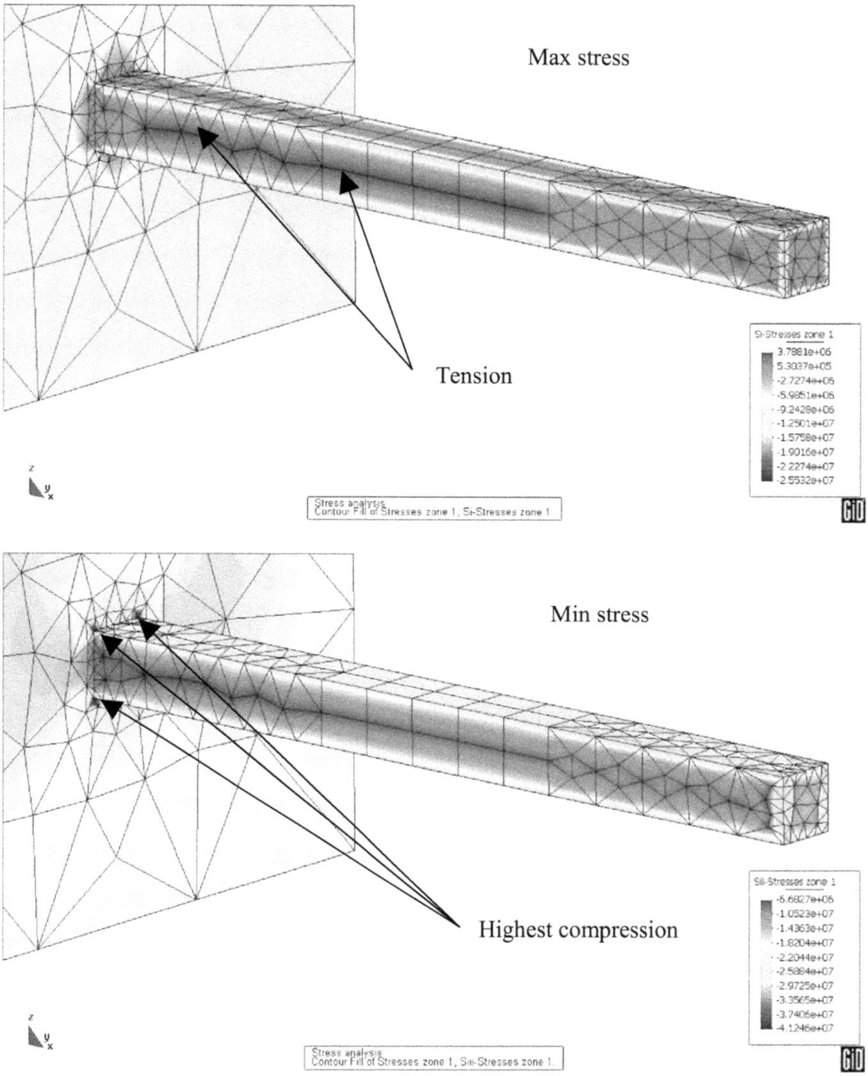

Figure 5.47: Maximum and minimum stress distribution in the submodel (20 MPa tectonic stress). Here the end side of the tunnel is visible.

In this last model the damage extends to a major part of the geometry considered, as shown in Figs 5.55 and 5.56. It is interesting to note that the pattern and extension of the failing where the fracture zone crosses the tunnel is very similar to the experiment with the same tectonic stress distribution (compare Fig. 5.55 on the left with Fig. 5.13). This confirms the validity of the two results and suggests that the influence of the room is very little already at a distance of 50 m.

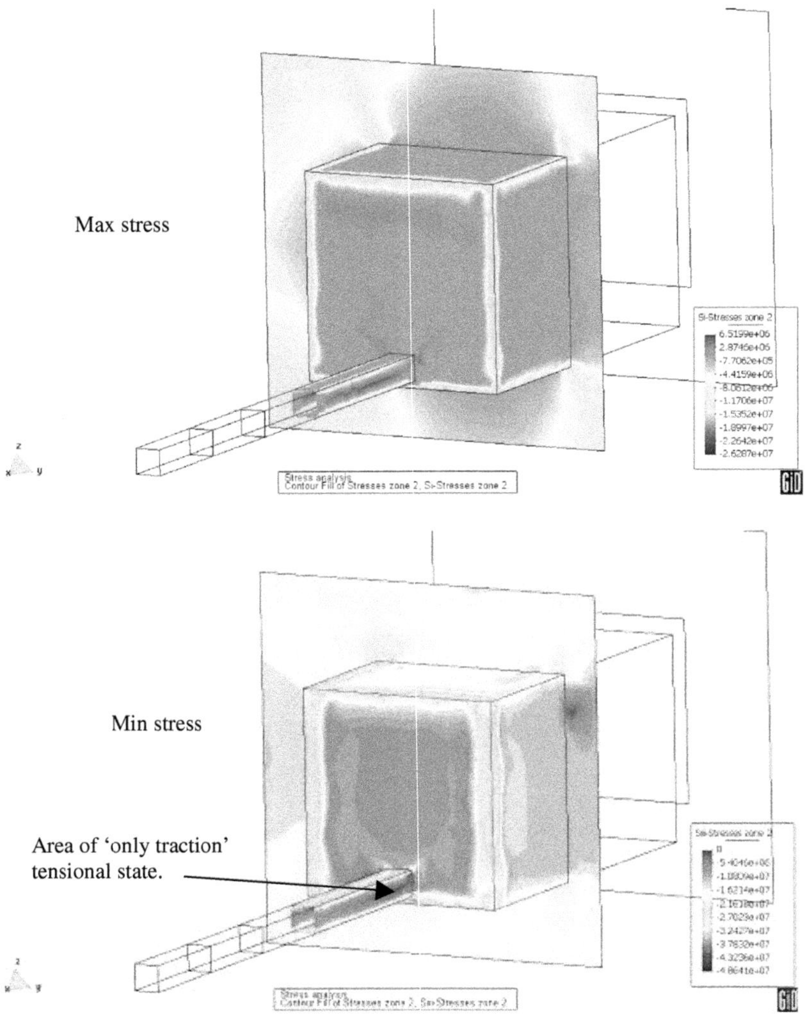

Figure 5.48: Maximum and minimum stress distribution in the submodel (20 MPa tectonic stress). Here the room entrance side is visible.

On the other hand, since on the room side (only) the damage reaches even the outer skin of the submodel, the hypothesis of submodelling becomes invalid, since the deformation and forces in the submodel will cause changes to the deformation and forces in the large model. Consequently a larger region must be used for the local submodel of the mine. Moreover, the extent of the failing rock probably means that under these conditions the whole block surrounding the room would collapse.

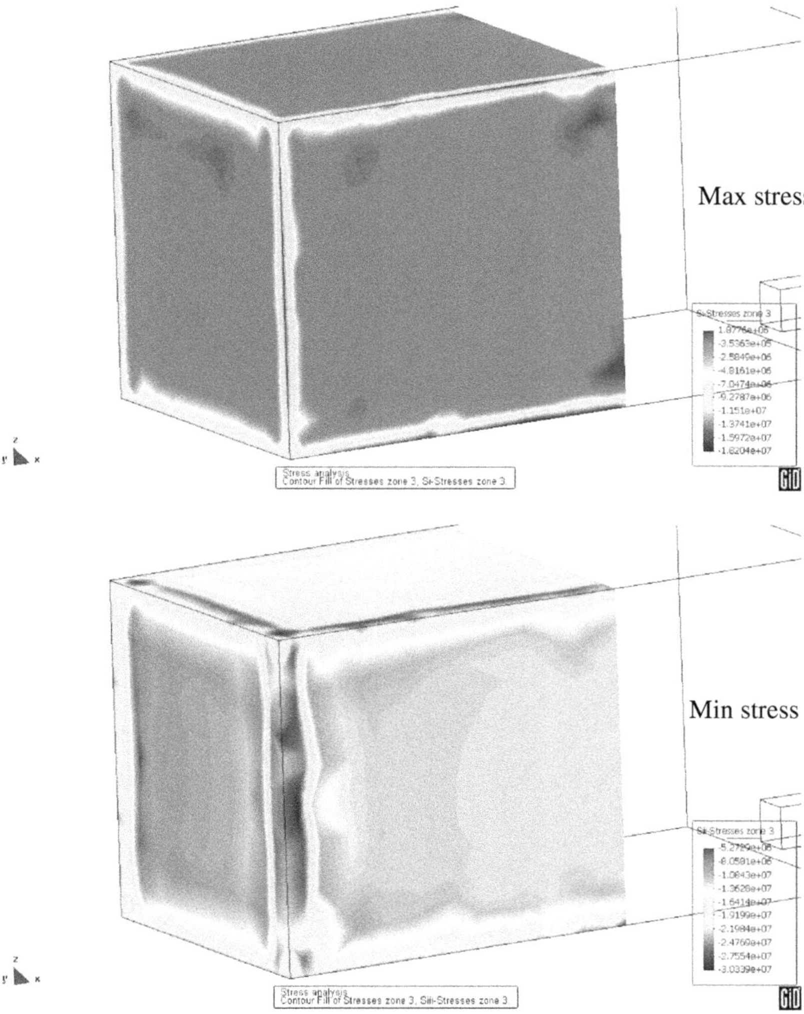

Figure 5.49: Maximum and minimum stress distribution in the submodel (20 MPa tectonic stress). Here the end wall of the room is visible.

5.6 Conclusions

Tools have been developed to investigate the stability of potential mine repository sites. The tools were based on the BEASY analysis software and the GiD modelling and visualization. Combining the facilities of BEASY for modelling fractured structures and extensive customization of the GiD program, a convenient tool has been developed to model the stability of mine repository sites in highly fractured materials like crystalline rock.

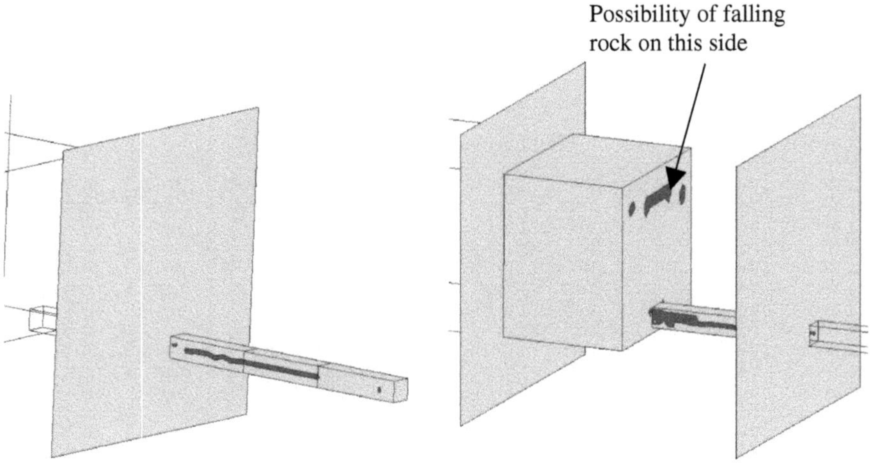

Figure 5.50: Mohr–Coulomb criterion on the tunnel and the front part of the room.

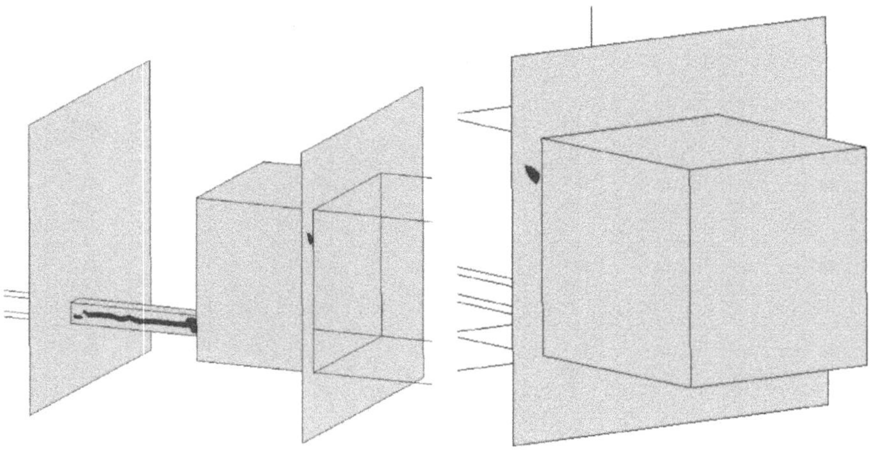

Figure 5.51: Mohr–Coulomb criterion around the room and in the first part of the tunnel.

Two reference mines have been investigated. The first is a typical limestone mine with room and pillar geometry. Two models were prepared and solved, representing the middle of the array and the end of the mine. The results appear to be consistent with the experience in the field. The second case was a mine in a crystalline rock with a fractured matrix (100 m distance between the fracture zones). Experiments on small detailed models showed that there was no practically significant influence of the EDZ on the rock stability. Therefore it was concluded that there is no need in this case to model the EDZ when assessing the stability.

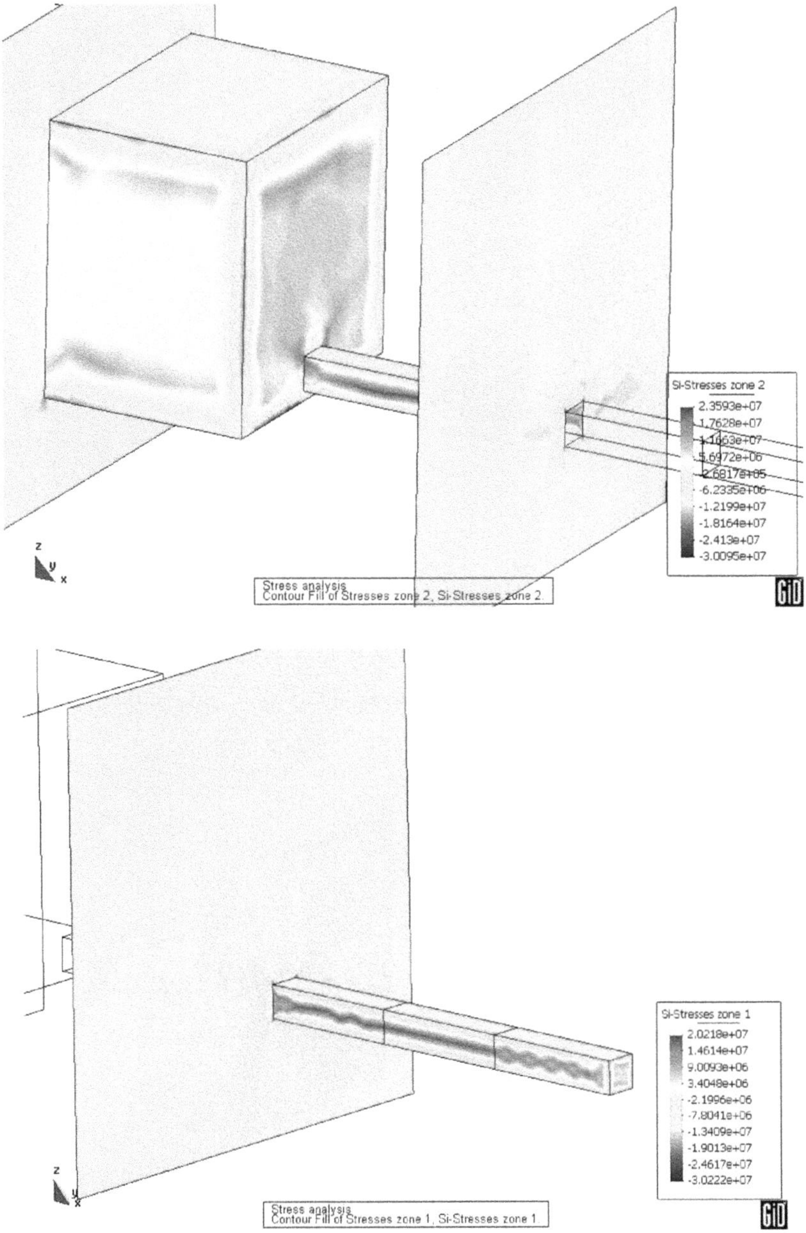

Figure 5.52: Maximum principal stress at the 'entrance side' of the room and along the tunnel in the case of 30 MPa tectonic pressure along room and tunnel.

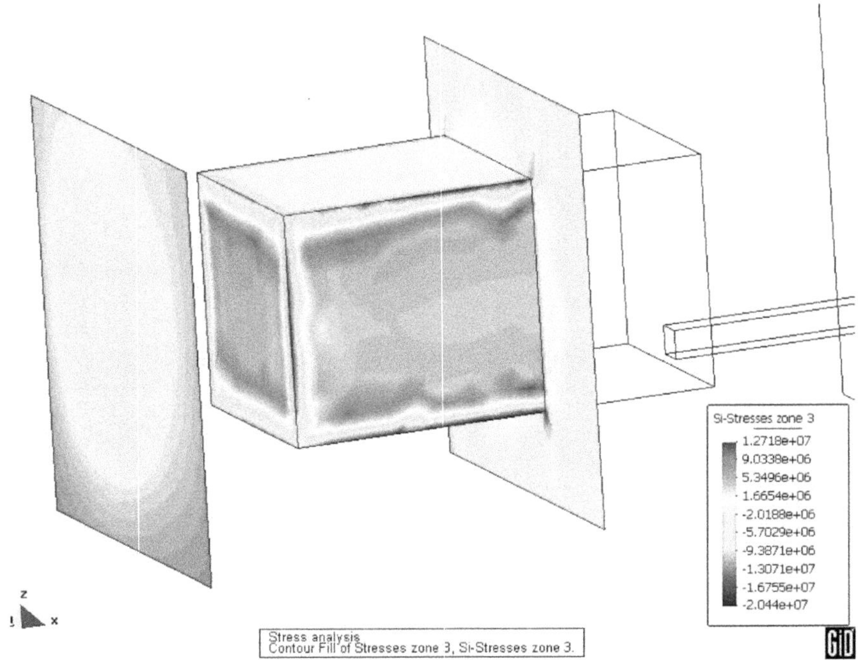

Figure 5.53: Maximum principal stress in the 'rear' side of the room.

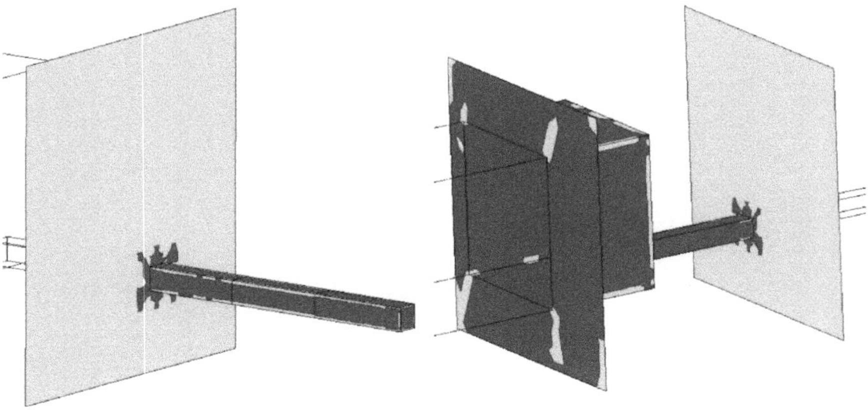

Figure 5.54: Mohr–Coulomb criterion on the tunnel and in the front part of the room.

It was concluded that a submodelling approach could be used in many of the cases. Software was therefore developed to map results and data from the large 'regional' based model to the smaller but much more detailed local model of the mine. The big 'regional' model was analysed first with the major tectonic boundary conditions applied. Then the results were transformed to the submodel,

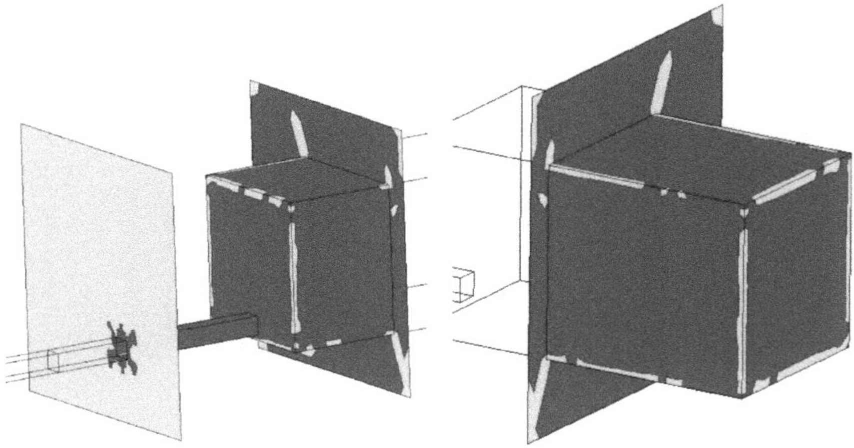

Figure 5.55: Mohr–Coulomb criterion on the room surfaces, tunnel and the fracture zone crossing the room.

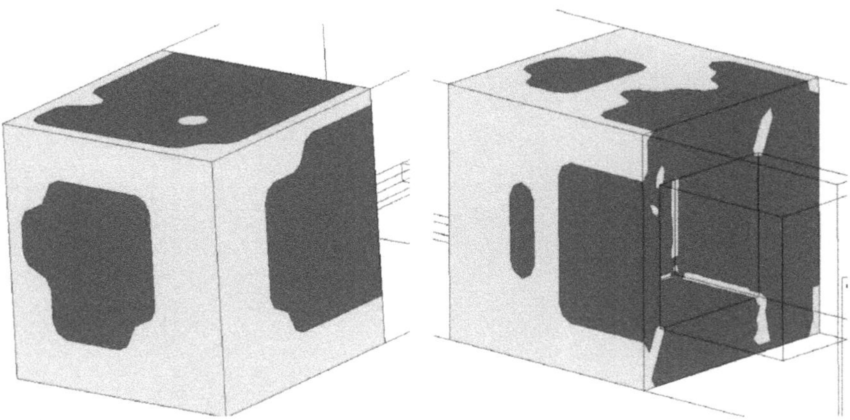

Figure 5.56: Mohr–Coulomb criterion on the outer side of the 'cubes' surrounding the room.

which was solved to reveal the detailed stress and deformation in the immediate vicinity of the mine.

For the crystalline rock case with tectonic stresses of the order of 20 MPa the stability analysis, performed using the Mohr–Coulomb failure criterion, shows that the tunnel and some part of the room are unstable. A second model with increased tectonic forces 30 MPa indicated substantial instability with the room totally damaged. With this configuration the mine at tectonic loads of 20 MPa or more would require strengthening.

For the limestone case the stress concentrates on the pillars starting from about 6–7 m above the ceiling. The Mohr–Coulomb failure extends all over the EDZ and in a thin layer above, making strengthening necessary.

CHAPTER 6

Risk assessment of underground repositories using numerical modelling of flow and transport in fractured rock

V. Popov & A. Peratta
Wessex Institute of Technology, Southampton, UK.

Abstract

This chapter describes the numerical modelling of flow and transport in fractured porous media, which was used in the research on the isolating capacity of underground repositories, and also presents the numerical results obtained for two cases: mine repository in limestone and mine repository in crystalline rock. It contains information on the governing equations of the model and the numerical technique applied for solving the equations. The waste-isolating capacity has been estimated for underground mines in two types of geological media: crystalline rock and limestone. The repository consists of a large room filled with hazardous waste embedded in clay, where in the case of crystalline host rock, there is also a tunnel extending from the room which is also used as a repository. In the case of granite there a number of fracture zones that intersect the domain and some of them intersect the excavation-disturbed zone. Two types of hazardous waste were considered: dichlorvos and batteries/zinc. The outline of this chapter is as follows: Section 1 introduces the background and a brief summary of the previous work in fractured porous media. Section 2 describes the governing equations for flow and transport in fractured porous media. Section 3 explains the numerical method, including the boundary element formulation, dual reciprocity method, and multidomain decomposition. Section 4 describes some features of the computational implementation. Section 5 describes the conceptual model proposed for the two cases considered: mine in fractured crystalline rock and mine in limestone, and shows the corresponding results in short and long-term calculations. Finally, Section 6 elaborates the conclusions and final remarks.

6.1 Overview of the problem

Understanding flow and transport processes in naturally fractured porous media is of interest in environmental engineering applications, in geohydrology or in oil reservoirs engineering, when porous strata are made of rocks, which are crossed by networks of fissures and cracks. Recently, fractured rocks attracted the attention in connection with the problem of geological isolation of hazardous waste. In particular, one immediate field of application is the conversion of abandoned mines into waste repositories, a technology that provides a feasible solution to the problem of hazardous waste management presenting many benefits from practical, technical and economical points of view.

The implementation of the proposed technology involves a number of safety and risk assessment considerations that mostly rely on field observations, laboratory tests and numerical modelling tools. The numerical modelling is a very useful tool for assessing the long-term safety of the repository. If one considers the complexity of the problems, including different geometries, many heterogeneous volumes, like for example various types of rock, fractures, clay, etc., all with variable characteristics, combined with sources, which vary in time, it would be impossible to obtain any form of flow or solute transport predictions, without using numerical modelling and computer simulations.

6.1.1 Scope and objectives

In general, the numerical modelling covers a variety of aspects including the definition of the conceptual model, the hypothesis and assumptions for the physical representation, the statement of the governing equations, the numerical methods, the computational implementation, and the calibration and validation of the results. In particular, this work is focused on the numerical method and solution of the governing equations, referring for the other aspects of the theoretical developments of the applied model to existing publications [1].

The underlying physical situation involves flow and transport phenomena in complex geological systems. The level of complexity is mainly defined by the geometrical and physical aspects of the model to be considered. The main objective in this chapter is to present a new numerical technique based on the coupling of 3D rock matrix and a fracture network based on 2D entities, solved using the boundary element method (BEM) implemented in a flexible domain decomposition way. The whole approach was developed in order to simulate the processes of flow and transient solute transport in 3D fractured porous media under the assumptions of the discrete fracture network model. The following sections of this chapter introduce a general overview of the fractured porous media concepts, which are necessary in order to implement the numerical model.

6.1.2 Fractured porous media

A porous medium can be regarded as a multiphase material body in which a representative elementary volume (REV) can be determined, such that the averaged hydraulic (permeability) and transport (dispersivity) properties of the domain are the same in each REV regardless of its position within the domain of interest. This is so because the size of the REV is sufficiently large that the physical parameters that represent the distributions of the void space and of the solid matrix within it are statistically meaningful [2].

On the other hand, a fractured porous medium (FPM) can be regarded as a composition of three well distinguished objects, namely individual fractures, fracture networks and the solid medium existing between the fractures (named also the porous matrix). A thorough description of FPM and fracture networks can be found in Adler and Thovert [3].

Individual fractures look like plane discontinuities when viewed from afar, or like two solid surfaces that surround an interstitial 3D space when viewed from close up. Fractures are regarded as porous media with usually higher permeability than the adjacent porous matrix and with apertures of many orders of magnitude smaller than the integration domain.

6.1.3 Overview

At present there are many different approaches for modelling flow and solute transport in FPM. One possible way to classify most of the available models is in terms of the degree of detail and accuracy that each one of them can describe according to their length ranges and space scales.

The two major divisions that appear naturally are basically the *microscopic representation*, in which the scale is such that it is possible to distinguish the void region inside a pore and the different phases within it; and the *macroscopic representation*, in which transport properties are usually averaged over REVs of porous material relatively large in comparison with the pore size, but at the same time small enough to describe local properties in the physical domain.

6.1.3.1 The continuum approach

The macroscopic representation yields to the continuum approach. Hence, under the continuum assumptions and according to the scale of the problem, there are four major subdivisions, as defined by Bear and Berkowitz [2].

6.1.3.2 The very near field zone

In which the region of interest is usually focused on a single well-defined fracture, embedded in the porous matrix.

6.1.3.3 The near field flow

In which a small amount of fractures, that may or may not intercept each other, defines a small and bounded fracture network, which is deterministically known.

This model allows statistical representation of the fractures with random shapes and orientations. Moreover, the fractures and the porous matrix represent well-defined separated regions in space.

6.1.3.4 The far field model

In this case, transport may be regarded as taking place in two overlapping continua. One for the fluid in the fracture network and the other for the porous matrix. There are material fluxes exchanged between both media that occupy the same physical location.

6.1.3.5 The very far field model

The fluid in the FPM as seen from afar may be regarded as a single continuum, with an equivalent permeability. The properties of the void space reflect both the presence of the fracture network and the porous matrix blocks.

This kind of model is applicable in cases where the system under consideration allows sufficient interaction between fluid and contaminants in the fracture and in the porous blocks, bringing the two systems to a local equilibrium at every (macroscopic) point. A model analogous to that of a regular porous medium can describe such a system.

6.1.3.6 The discrete fracture model

The discrete fracture network model adopted in this work can be regarded as a network of interconnected fractures embedded in a porous matrix. Each individual fracture is represented by an arbitrary surface, that may be plane or not, finite or infinite and of various shapes.

In general, the main feature of the fractures is that they might cover a wide range of scales, from submillimetric fissures to long faults of hundred of kilometres. Another distinctive feature is that fractures might intersect randomly each other, generating a complex interconnected network.

The goal of this work is to model the fractures individually, in such a way that they are deterministically described so that the exact location and geometry are known and predefined.

When viewed from a close distance, each surface of the fracture network can be represented as a 3D object with one of its dimensions, the aperture, of several orders of magnitude smaller than the other two. The aperture of the individual fracture, namely w_f, might vary from point to point within the fracture. A convenient simplification of the model is to decompose each individual surface of the fracture network into smaller fracture elements. Each element has associated a constant aperture that might be different from the others. In this way a piecewise varying field of aperture can be prescribed over every single surface of the network.

Although each fracture has three dimensions, the model proposed in this work regards the aperture as a scalar field attached to a 2D surface built from a cluster of interconnected flat fracture elements, thus being represented by two local coordinates.

The fracture network is itself made of porous media, usually of higher permeability than the adjacent porous matrix in which it is embedded.

For the purpose of this work, it is enough to calculate the permeability of the fracture network by means of the Poiseuille law or by data collected from experimental measurements.

6.1.4 Historical development of porous media modelling

A very important contribution that established the starting point of modern modelling in this topic is the dual-porosity model of Barenblatt *et al.* [4] who represented the fractured porous media as a continuum consisting of two overlapping regions/continuums, porous matrix and fracture network, where the coupling term between the two pore systems was derived, among the others, by Gerke and van Genuchten [5].

In the field of deterministic fracture networks, Warren and Root [6] gave a very important contribution with orthogonal structured fracture networks. Later Odeh [7] presented a generalization of this work with arbitrary unstructured fracture network patterns. In the range of local scale modelling, it is worth mentioning the contribution of Snow [8] where he found exact expressions of permeabilities for infinite fractures. Brown and Scholz [9] and later Gentier [10] introduced a new method to characterize systematically natural fracture networks.

This characterization was subsequently used by Brown [11] and Moreno *et al.* [12, 13] to determine the permeability of a single fracture by integration of the 2D Reynolds equation. Barton *et al.* [14] compiled many years of research on hydromechanical joint properties in a coupled joint behaviour model.

Lately, Mourzenko *et al.* [15] reformulated the calculation of single fracture permeabilities by integrating the Stokes equations. On the fracture network scale, significant efforts where made to model transport properties and to match the numerical results with those obtained from the experiment. Bond networks were extensively used by Dienes [16], Long *et al.* [17] and Cacas *et al.* [18] with an elegant method to generate off-lattice bond networks in three dimensions.

Mercer and Faust [19] have demonstrated the circumstances in which the very far field approach can be used to describe flow and heat transport in porous media. During the eighties the trend was the application of concepts derived from the *percolation theory* for the study of fractures. Percolation deals with the effects of varying the richness of interconnections present in a random system. The basic idea of percolation is the existence of a sharp transition at which the long-range connectivity of the system disappears (or, going the other way, appears). This transition occurs abruptly when some generalized density in this system reaches a critical value called *percolation threshold.*

Researchers attempted to determine the percolation threshold, in this case interpreted as the density of fractures above which the connectivity of fracturesis sufficient to enable flow through the network, or at least through part of the fractures. Dienes [16] and Charlaix *et al.* [20] provided key contributions in this subject bringing new light in network analysis. Also Robinson [21], Charlaix *et al.* [22] and Wilke *et al.* [23] studied the issue of connectivity with percolation theory.

An innovative way to deal with fractures as interfaces in porous blocks has been implemented and tested in two dimensions (the porous matrix is represented in two dimensions, while the fractures are represented as 1D curves) with the finite element method [24]. The strategy employed in this work is somewhat similar, with the main difference that the implementation is in 3D space with the fractures being 2D entities.

6.2 Governing equations

The mathematical formulation, using discrete fracture model, is divided into two major parts, namely flow and solute transport in matrix block and fracture network. The former is concerned about the velocity field and the hydraulic head in the FPM, whereas the latter is concerned with the concentration, and the concentration flux fields of contaminants.

Based on the hypothesis of low concentrations and incompressibility of the medium, both parts of the model can be decoupled and solved in a sequential way. The flow problem is solved first in order to define the velocity field, and then the velocity field information is passed to the numerical model solving the solute transport in order to define the convective term.

6.2.1 Flow

6.2.1.1 General formulation

The flow model in porous media is based on the continuum approach, and it is described by:

$$\nabla \cdot (k \nabla h) + Q_e = S_r \frac{\partial h}{\partial t}, \tag{6.1}$$

which was derived under the following assumptions: isothermal conditions, homogeneous fluid, hydraulic conductivity independent of pressure changes, and specific storativity and hydraulic conductivity unaffected by variations of the porosity. The hydraulic head is calculated according to:

$$h = \frac{p}{\rho g} + z, \tag{6.2}$$

where S_r represents the specific storativity:

$$S_r = \rho g (\alpha + \beta \vartheta), \tag{6.3}$$

α is the rock matrix compressibility, given by:

$$\alpha = \frac{1}{1-\theta}\frac{\partial \theta}{\partial p} \tag{6.4}$$

and β is the compressibility of the fluid:

$$\beta = \frac{1}{\rho}\frac{\partial \rho}{\partial p}, \tag{6.5}$$

Where θ is porosity of the medium, p is the pressure, ρ is the density of the solution, g is the gravity acceleration, z is the coordinate along direction of gravity, and Q_e represents possible sources or sinks.

Equation (6.6) represents the conservation of mass according to the Darcy law which relates the velocity v to the gradient of hydraulic head:

$$v = -k\nabla h. \tag{6.6}$$

6.2.1.2 Flow in the porous matrix

In particular, when the characteristic diffusion time defined as $\tau_D = L^2 S_r/k$ where L is the length scale of the problem, is several orders of magnitude smaller than the modelling time (time scale of practical interest), it is reasonable to neglect the time-dependent term and to consider a steady state solution for the flow. (This is a key hypothesis in the present chapter: in groundwater flow the compressibility factors are small and the conductivity high, such that $\tau d >> t$.) Under those circumstances, it is a good approximation to consider the potential equation for the hydraulic head in which case, the governing equation in the porous matrix becomes

$$\nabla \cdot \left(-k_m \nabla h_m\right) = 0. \tag{6.7}$$

6.2.1.3 Flow in a single fracture

Let us consider the tangential plane to the surface that represents the fracture, and a local system of coordinates (x', y', z') such that z' is coincident with the local normal to the plane. The discrete fracture network model assumes that

$$\begin{aligned} w_f &\gg L_f, \\ k_f &\gg k_m, \end{aligned} \tag{6.8}$$

where w_f is the local aperture of the fracture and L_f is the extension of the fracture. Therefore, it is possible to represent the solution in the fracture as a superposition of the 2D solution on the tangential plane and the 1D profile $f(z)$ along z'

$$h(x, y, z) = h_f(x', y'), \quad 0 \le z' \le w_f. \tag{6.9}$$

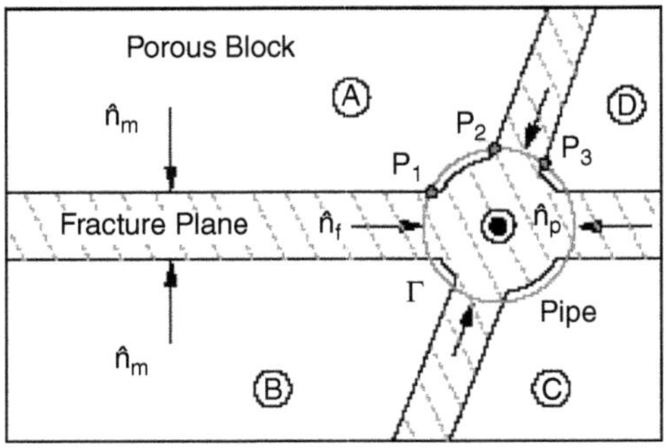

Figure 6.1: Intersection of two fractures.

Integration of (6.7) along z' yields

$$-k_f \overline{w}_f \nabla^2_{2\mathrm{D}} h_f = \sum_{i=A,B} v_{m,i}, \tag{6.10}$$

where

$$\nabla_{2\mathrm{D}} := \nabla(...) - \nabla() \cdot \mathbf{z}'. \tag{6.11}$$

The indices A and B stand for the two blocks of porous matrix adjacent to the fracture. v_m is the velocity in the porous matrix boundary that limits with the fracture pointing parallel to its outward normal $\mathbf{n}_m$ (see Fig. 6.1); and w_f is the equivalent aperture of the fracture.

$$\overline{w}_f = \int_0^{w_f} f(z')\,\mathrm{d}z' =: w_f(1+\beta). \tag{6.12}$$

6.2.1.4 Fracture intersections

In any arbitrary interconnected fracture network, a certain number (m_f) of fractures might intercept each other converging into a common channel. The resulting channel might have material properties significantly different from its adjacent environment (either fractures or porous matrix), in the same way that fractures represent a discontinuity for the adjacent blocks of porous matrix. Furthermore, regarding that single fractures are represented by surfaces, it is natural to represent their intersections by 1D curves in the 3D space. In the fracture intersections or *pipes* (the theoretical representation of the real *fracture intersections* or channels), integration of the continuity equation over the cross section A_p (see Fig. 6.1.) yields:

$$\int_{A_p} -k_p \frac{\partial^2 h_p}{\partial \eta^2}\mathrm{d}A = \int_{\text{``}} v_{fn}\,\mathrm{d}\Gamma, \tag{6.13}$$

where

$$v_{fn} = -k\frac{\partial h_f}{\partial n_f} \tag{6.14}$$

is the velocity along the fracture planes adjacent to the pipe in the direction of the outward-to-the-fracture normal unitary vector perpendicular to the longitudinal axis of the pipe $\hat{n}_f$, and η is the coordinate along the pipe (unit vector $\hat{n}_p$, see Fig. 6.1). Assuming that.

1. the main contribution of the flux in the right-hand side integral of (6.13) comes from the adjacent fractures, thus neglecting the part of the integral involving the contact between the lateral surface of the pipe and the porous matrix, and that
2. the hydraulic head in a given cross section is constant (it depends only on η, the local coordinate along the pipe), then (6.13) can be expressed in the following way

$$-A_p k_p \frac{\partial^2 h_p}{\partial \eta^2} = \sum_{i=1}^{m_f} w_{f,i} v_{f,n,i}, \tag{6.15}$$

where i labels each adjacent fracture element to the pipe.

6.2.1.5 Flow in pipe connectors

In the same way that fractures intersect each other creating channels, an arbitrary number of channels might intersect each other creating pipe connectors. These objects can be regarded as closed volumes of similar extension in all directions and comparable with the mean diameter of all the convergent channels $\sqrt{A_p}$. By analogy, it is consistent to represent channel connectors by points called multiple pipe connectors (MPCs), disregarding their 3D structure by integration in volume. Thus, in an MPC, the following 0D version of the mass conservation is considered for the flow:

$$\sum_{i=1}^{m_p} A_{pi} \cdot v_{\eta,i} = 0, \tag{6.16}$$

together with the continuity of hydraulic heads

$$h_{p,1} = h_{p,2} = \ldots = h_{p,m_p}, \tag{6.17}$$

where m_p is the number of pipes converging to the point.

6.2.2 Transport

This section presents the governing equations in the porous matrix, a single fracture, a single pipe, and a single MPC. It is considered that all these entities offer the same kind of discontinuity for both flow and transport problems, so the derivation in this case is similar to the one for the flow, being the only difference the leading operator of the partial differential equation (PDE). Finally, the formulation for the complete problem where all the entities interact at the same time is the result of solving all the equations together. The description of the coupling strategies will be treated in the next two sections.

6.2.2.1 General formulation

In general, the transport process under consideration can be described by the advection–diffusion reaction equation for the concentration of pollutant:

$$R\frac{\partial c}{\partial t} + \nabla \cdot q = K_r c + \rho, \tag{6.18}$$

$$q_i = v_i c - D_{ij}\frac{\partial c}{\partial x_j}, \tag{6.19}$$

$$D_{ij} = D^{\mathrm{M}} + \alpha_{\mathrm{L}} P_{ij}(v) + \alpha_{\mathrm{T}}\left(\delta_{ij} - P_{ij}(v)\right)v, \tag{6.20}$$

where R is retardation factor, q is solute flux, K_r is reaction constant, ρ is a source, D_{ij} is the dispersion coefficient; D^{M} is the molecular diffusivity; α_{L} and α_{T} are the longitudinal and transversal dispersitivies, respectively, and $P(\boldsymbol{v})$ is the projection operator onto the direction of the velocity vector $\boldsymbol{v}$.

$$\boldsymbol{v} = \boldsymbol{v}\left(l_x\hat{x} + l_y\hat{y} + l_z\hat{z}\right) \tag{6.21}$$

$$P(\boldsymbol{v}) = \begin{bmatrix} l_x l_x & l_x l_y & l_x l_z \\ l_y l_x & l_y l_y & l_y l_z \\ l_z l_x & l_z l_y & l_z l_z \end{bmatrix}. \tag{6.22}$$

In order to simplify the notation in the following sections, it is practical to define the p-dimensional advection–diffusion reaction operator L_{χ_p} applied on the entity χ by:

$$L_{\psi_p} := R_\psi\frac{\partial u}{\partial t} + v_{\psi,i}\frac{\partial}{\partial x_i} - D_\psi\frac{\partial^2}{\partial x_i^{\,2}} - k_r, \tag{6.23}$$

where the subindex χ can be any of m, n or p identifying the porous matrix, a single fracture, or a pipe element , respectively; the index p is the dimensionality (1, 2 or 3); and $i = 1,\ldots, p$.

Then, the general formulation for the transport in any entity of the FPM can be summarized in the following expression:

$$L_\chi^p[c_\chi] = \rho_\chi, \tag{6.24}$$

Table 6.1: Entities involved in the problem and their possible dimensionality.

		Dimension p	
Entity	χ	3D case	2D case
Porous matrix	m	3	2
Fracture	f	2	1
Pipe	p	1	0
MPC	o	0	–

where ρ_χ changes according to χ and p, and represents the coupling term between entities. Table 6.1 summarizes the possibilities.

6.2.2.2 Transport in the porous matrix

Both, the small variation of concentration and the assumption of low compressibility of the fluid allow us to neglect any change of the density, and to keep the approximation of the irrotational steady state velocity field. In addition, considering a homogeneous isotropic media, the concentration in the porous matrix is described by (6.24) with $\chi = m$ and $\rho_m = 0$. (For simplicity, the theory is presented for homogeneous properties, in spite of the fact that the numerical method explained in the next two sections, has been developed for non-homogeneous piecewise constant diffusion coefficients.)

6.2.2.3 Transport in a single fracture

Integration of (6.19) along z' in the same way as was done in the flow problem yields

$$\rho_f = \frac{1}{\overline{w}_f} \sum_{i=A,B} q_{m,n,i}, \tag{6.25}$$

where $q_{m,n,i} = \mathbf{q} \cdot \hat{n}_{m,i}$ is the normal concentration flux coming from the porous matrix. Thus, the influence of the two blocks of porous matrix (A and B) adjacent to the fracture is considered as a source term for the fracture.

Finally, the formulation for a single fracture is represented by (6.24) and (6.25) with $\upsilon = f$.

6.2.2.4 Transport in pipes

Integration of (6.18) over the cross section of a channel yields the 1D formulation for pipes, represented by (6.24) with $\chi = p$ and the source term given by:

$$\rho_p(\eta,t) = \frac{1}{A_p} \int_\Gamma v_{f,n} c_f(\eta,s,t) - D_f \nabla_{2\mathrm{D}} c_f(\eta,s,t) \cdot \hat{n} \mathrm{d}\Gamma \tag{6.26}$$

being η the coordinate along the pipe.

6.2.2.5 Transport in pipe connectors

Similarly to the flow problem, the 3D structure of an MPC is collapsed into a point where L_o^0 is identically zero, and the source term of (6.24) becomes:

$$\rho_o(\eta,t) = \int_{\partial V_o} q_{p,x} \cdot \hat{n} \mathrm{d}S_o. \tag{6.27}$$

Thus leading to the continuity of concentration

$$c_{p,1} = c_{p,2} = \ldots = c_{p,m_p} \tag{6.28}$$

and the conservation of the normal flux of concentration

$$\sum_{i=1}^{m_p} A_p q_{p,n,i} = 0. \tag{6.29}$$

6.3 Numerical method

6.3.1 Introduction

This section presents a general overview of the numerical scheme to be implemented. In Section 6.3.2 the main steps of the BEM are summarized, a complete introduction to the method can be found in Brebbia *et al.* [25].

Section 6.3.3 describes the dual reciprocity method (DRM) which is used to solve the domain integrals that appear in the integral formulation of the governing equations. Finally, the third part describes the time integration scheme and the domain decomposition technique.

6.3.2 The boundary element method

6.3.2.1 Integral formulation

The numerical implementation of the BEM requires the discretization of the boundary into elements, and this represents one of the most powerful advantages of the method, since there is no need to discretize in volume.

The starting point of the BEM is the integral formulation of the governing differential equation. The governing equation that describes a linear time-dependent process of flow and transport in porous media, in a general form in a domain Ω, can be written as

$$\nabla^2 u = b(x, y, u, t), \tag{6.30}$$

where the boundary conditions are defined as

$$u = \bar{u} \quad \text{on } \Gamma_1 \tag{6.31}$$

and

$$q = \partial u / \partial n = \bar{q} \quad \text{on } \Gamma_2. \tag{6.32}$$

Here $\Gamma = \Gamma_1 + \Gamma_2$ is the exterior boundary that encloses the domain Ω and n is its outward normal.

For the flow equation, u represents the hydraulic head h and therefore according to (6.1) the non-homogeneous term b has the following form:

$$b = \frac{1}{k}\left(-Q_e + S_r \frac{\partial h}{\partial t}\right), \tag{6.33}$$

while for the transport, $u \equiv c(x_i, t)$ and according to (6.18) and (6.19) for scalar dispersion coefficient the non-homogeneous term b has the following form:

$$b = \frac{1}{D}\left(R\frac{\partial c}{\partial t} + \nabla(\mathrm{v}_i c) - K_r c - \rho\right), \tag{6.34}$$

where $v_i = -k(\partial h / \partial x_i)$ represents the ith component of the velocity vector v shown in (6.6).

Applying the Green integral representation formula to (6.30), it is found that the value u at a point x within the domain Ω, is given by [25]:

$$\lambda(x)u(x) + \int_\Gamma q^*(x,y)u(y)\mathrm{d}\Gamma_y - \int_\Gamma u^*(x,y)q(y)\mathrm{d}\Gamma_y = \int_\Omega u^*(x,y)b(y)\mathrm{d}\Omega_y \tag{6.35}$$

Here, $u^*(x, y)$ is the fundamental solution of the Laplace equation, which for an isotropic 2D medium is given as:

$$u^*(x,y) = \frac{1}{2\pi}\log\frac{1}{r}, \tag{6.36}$$

and for a 3D case becomes:

$$u^*(x,y) = \frac{1}{4\pi}\frac{1}{r}, \tag{6.37}$$

where r is the distance from the point of application of the concentrated unit source to any other point under consideration, i.e. $r = |x - y|$, $q(y) = \partial u(y)/\partial n$ and $q^*(x, y) = \partial u^*(x, y)/\partial n$ and n is the normal to the boundary. Note that in (6.35) all the integrals are over the boundary Γ of the domain, therefore these are surface integrals, except for the one corresponding to the term $b(y)$, which is performed over the domain Ω, and therefore this is a volume integral. The term $b(y)$ represents the sum of the non-homogeneous terms, see (6.33) and (6.34), and is defined according to the model and process under consideration. The constant $\lambda(x)$ has values between 0 and 1, where for a smooth boundary it is equal to 1/2, and for points inside the domain $\lambda(x)$ is equal 1. More information on how to calculate $\lambda(x)$ can be found in Brebbia *et al.* [25].

Equation (6.35) represents the integral formulation of (6.30), and is the starting point of the applied BEM formulation.

6.3.2.2 Boundary discretization

The three terms on the left-hand side of (6.35) involve only boundary integrals. The boundary Γ can be discretized into N_{e3D} elements. In the 3D case, the integration domain Ω (used to represent the porous matrix blocks) is a volume, and its boundary Γ is discretized by means of a collection of connected triangular or quadrilateral isoparametric elements. In the 2D case, the boundary is represented by lines and the integration domain Ω is a surface.

The path enclosing Ω is the boundary, which is discretized into N_{e2D} linear elements. The BEM for the 1D case will be used to represent pipes when intersections between fractures in the fracture network occur. The integration domain (Ω) is a linear straight segment and their two geometrical endpoints become the boundary (Γ), and each one of them represents a boundary element.

In general, Γ is discretized into N_e boundary elements, according to: $\Gamma = \Gamma_1 \cup \Gamma_2 \cup \ldots \cup \Gamma_{N_e}$, so that

$$c_i u_i + \sum_{j=1}^{N_e} \int_{\Gamma_j} \frac{\partial u^*}{\partial n} u \, \mathrm{d}\Gamma - \sum_{j=1}^{N_e} \int_{\Gamma_j} u^* \frac{\partial u}{\partial n} \mathrm{d}\Gamma + \int_{6} b u^* \mathrm{d}\Omega = 0. \tag{6.38}$$

The treatment of the domain integral that appears in the last term of (6.38) will be deferred for the next section. Each boundary element contains a number N_{fn} of subjacent collocation nodes, where the potential or fluxes are evaluated. In this way, the values of the potential or its normal derivative at any point defined by the local coordinates on a given boundary element can be defined in terms of their values at the collocation nodes, and the N_{fn} interpolation functions in the following way:

$$u(\xi) = \sum_{k=1}^{N_{fn}} \psi_k(\xi) u_k \tag{6.39}$$

and

$$\frac{\partial u(\chi)}{\partial n} = \sum_{k=1}^{N_{fn}} \upsilon_k(\chi) \left.\frac{\partial u}{\partial n}\right|_k . \tag{6.40}$$

With the discretization of the boundary and using the collocation technique, expression (6.38) can be rewritten in the following way:

$$c_i u_i + \sum_{j=1}^{N_e} \sum_{k=1}^{N_{fn}} \left(\int_{\Gamma_j} \frac{\partial u^*}{\partial n_j} \theta_k \mathrm{d}\Gamma_j \right) u_{kj}$$

$$- \sum_{j=1}^{N_e} \sum_{k=1}^{N_{fn}} \left(\int_{\Gamma_j} u^* \theta_k \mathrm{d}\Gamma_j \right) \left. \frac{\partial u}{\partial n} \right|_{kj} + \int_{6} b u^* \mathrm{d}\Omega = 0. \tag{6.41}$$

The notation can be simplified by making use of matrix notation, so the last expression can be written in the following way:

$$\mathbf{Hu} - \mathbf{G} \left[\frac{\partial \mathbf{u}}{\partial n} \right] = - \int_{6} b u^* \mathrm{d}\Omega, \tag{6.42}$$

where

$$H_{il} = \delta_{il} c_i + \int_{\Gamma_j} \left. \frac{\partial u_i^*}{\partial n_j} \right|_{\chi_j} \upsilon_k(\chi_j) \mathrm{d}\Gamma_j, \tag{6.43}$$

and

$$G_{il} = \int_{\Gamma_j} u_j^*(\xi_j) \psi_k(\xi_j) \mathrm{d}\Gamma_j, \tag{6.44}$$

where the index $l = 1, \ldots, N_{fe}$ and $N_{fe} = \sum_{j=1}^{N_e} N_{fn,j}$ is the total number of collocation nodes adjacent to a given domain.

In fact, the index l is used to identify one of the adjacent freedom (collocation) nodes from a global point of view, and is given as a function of the indicator of element (j), and the local collocation node of that element (k). The boundary element $\mathrm{d}\Gamma_j$ can be expressed in terms of the domain local coordinates (ξ) through the Jacobian of the transformation $|J|$ in the following way:

$$\mathrm{d}\Gamma_j = |J| \mathrm{d}\xi_1 \cdots \mathrm{d}\xi_h, \tag{6.45}$$

where h is the dimension of Γ. Finally, provided that the right-hand side term of eqn (6.42) can be written as a given vector in function of the source term, or a characteristic matrix in function of the unknown potentials and normal fluxes at the collocation nodes of the boundary, the application of the prescribed boundary conditions and the assembly of the linear set of equations, that (6.42) produces, yields to a determined system of equations of dimension $N_{fe} \times N_{fe}$ of the form

$$\mathbf{Ax} = \mathbf{b}, \tag{6.46}$$

where the vector of unknowns (**x**) contains the potentials and normal fluxes that were not prescribed as boundary conditions. The matrix **A** contains the coefficients of **H** and **G**, and the right-hand side term contains the non-homogeneous term and the boundary conditions.

6.3.2.3 Internal solution

Once the problem given by (6.46) is solved, it is possible to calculate the values of the fluxes $\nabla u(\mathbf{x}_i)$ and potentials $u(\mathbf{x}_i)$ at any internal (observation) node $\mathbf{x}_i$ by means of the integral eqn (6.41). Therefore, the potential at $\mathbf{x}_i$ becomes:

$$u_i = \sum_{j=1}^{N_{fe}} \left[\frac{\partial u_j}{\partial n} \right] G_{ij} - \sum_{j=1}^{N_{fe}} u_j H_{ij} \tag{6.47}$$

and the gradient of the potential can be obtained with:

$$\left. \frac{\partial u}{\partial x_p} \right|_{x_i} = \sum_{j=1}^{N_{fe}} \int_{\Gamma_j} \frac{\partial u_j}{\partial n} \frac{\partial u^*}{\partial x_p} \mathrm{d}\Gamma_j - \sum_{j=1}^{N_{fe}} \int_{\Gamma_j} u_j \frac{\partial}{\partial x_p} \left(\frac{\partial u^*}{\partial x_p} \right) \mathrm{d}\Gamma_j. \tag{6.48}$$

6.3.3 The dual reciprocity method

This section gives an overview of DRM including the radial basis function considered, and the strategy for the reactive, advective, and time-dependent term. The previous section gave a general overview of the BEM for the Poisson equation, avoiding the treatment of domain integrals. In general, domain integrals arise from linear but non-homogeneous terms, non-linear terms, or time-dependent terms. In this case, the non-linear term in (6.42) introduces one of those domain integrals in (6.36). The most familiar techniques used to solve domain integrals are, direct numerical approximation, elimination of non-homogeneous terms via exact or approximate particular solutions, and dual and multiple reciprocity methods.

In principle, the domain integral would require some internal discretization in which case the complete scheme would loose one of its main attractions, which is being based on a 'boundary-only' discretization. Although internal discretization has been extensively used in the past, e.g. in the cell integration method [26], providing accurate results for a variety of PDEs, it has the main disadvantage of requiring an extra amount of data such as internal connectivities, hence making the code more complex and more demanding in terms of computational resources. Here, the DRM is proposed in order to avoid this inconvenience.

The DRM was introduced by Nardini and Brebbia [27] and subsequently used in various applications. A thorough introduction to the method can be found in Partridge *et al.* [28].

The main idea is to transform the domain integral that appears in (6.42) into an integral over the boundary by means of a finite set of interpolating functions, as explained in the next section.

6.3.3.1 General approach

The non-homogeneous term b in (6.30) can be written as a linear combination of the approximating functions f_j

$$b(\mathbf{x}) = \sum_{j=1}^{N_r} \alpha_j f_j(\mathbf{x}), \tag{6.49}$$

where N_r is the number of functions required for the approximation and α_j are undetermined coefficients. The approximating functions are linked to the particular solution $\hat{u}$ of the Laplace operator through

$$\nabla^2 \hat{u}_j = f_j. \tag{6.50}$$

Thus, eqn (6.30) can be written in the following way:

$$\nabla^2 \hat{u} = \sum_{j=1}^{N_r} \alpha_j \nabla^2 \hat{u}_j. \tag{6.51}$$

In the last expression it is possible to apply the weighting procedure with the fundamental solution in order to produce the integral equation

$$\int_\Omega \left(\nabla^2 u\right) u^* \mathrm{d}\Omega = \sum_{j=1}^{N_r} \alpha_j \int_\Omega \left(\nabla^2 \hat{u}_j\right) u^* \mathrm{d}\Omega. \tag{6.52}$$

Applying the Green integral representation formula and the subsequent discretization of the boundary, as it was described in the previous section, yields the following integral equation, for the source point located at the ith collocation node:

$$\begin{aligned} &\sum_{k=1}^{N_{fn}} H_{ik} u_k - \sum_{k=1}^{N_{fn}} G_{ik} \frac{\partial u_k}{\partial n} \\ &= \sum_{j=1}^{N_r} \alpha_j \left(\sum_{k=1}^{N_{fn}} H_{ik} \hat{u}_{kj} \right) - \left(\sum_{k=1}^{N_{fn}} G_{ik} \frac{\partial u_{kj}}{\partial n} \right), \end{aligned} \tag{6.53}$$

where

$$\hat{u}_{ij} = \hat{u}_j(\mathbf{x}_i) \tag{6.54}$$

is the jth particular solution evaluated at the ith collocation point, and

$$\frac{\partial \hat{u}_{kj}}{\partial n} = \frac{\partial \hat{u}_j(\mathbf{x}_k)}{\partial n_{[k]}} \tag{6.55}$$

is the derivative of the jth particular solution evaluated at kth collocation node in the direction of the outward normal of the boundary element that contains the kth node. The vector of coefficients (α_j) in eqn (6.53) can be obtained by solving the linear system:

$$\mathbf{F}\boldsymbol{\alpha} = \boldsymbol{b}, \tag{6.56}$$

where $\boldsymbol{b} = \boldsymbol{b}(\mathbf{x}_i)$ is the vector containing the non-homogeneous terms evaluated at the collocation nodes.

Finally, it is more practical to rewrite eqn (6.53) in the following matrix notation:

$$\mathbf{Hu} - \mathbf{G}\frac{\partial \mathbf{u}}{\partial \boldsymbol{n}} = \mathbf{S}\boldsymbol{b}, \tag{6.57}$$

where the following DRM matrices have been defined

$$\hat{\mathbf{U}} = \hat{u}_{ij}$$

$$\hat{\mathbf{Q}} = \frac{\partial \hat{u}_j(\mathbf{x}_k)}{\partial n_k}$$

$$\mathbf{S} = \left(\mathbf{H}\hat{\mathbf{U}} - \mathbf{G}\hat{\mathbf{Q}}\right)\mathbf{F}^{-1}. \tag{6.58}$$

6.3.3.2 Radial basis functions

In principle, any set of approximation functions f could be used in the DRM formulation. The only restriction is that the resulting matrix **F** must be non-singular. At the same time it is desirable to minimize high amplitude oscillations without excessively smoothing the interpolation. The interpolating functions used in this work for the transport problem in the fractures, and in the porous matrix were the so-called augmented thin plate splines [29], whereas the solution in the pipes, is based on cubic splines and Lagrange polynomials. Table 6.2 summarizes the sets of interpolating functions used in each case.

When an element of matrix **F** is represented by a radial basis function (RBF), then the following notation is equivalent:

$$f_{ij} = f_j(x_i) = f(r_{ij}), \tag{6.59}$$

where r_{ij} is the distance between the collocation nodes i and j.

Table 6.2: Sets of interpolating functions used.

Entity	Dimension	Set of interpolating functions
Porous matrix	3D	$\{r, 1, x, y, xy\}$
Fracture	2D	$\{r^2\log(r), 1, x, y\}$
Pipe	1D	Cubic splines
MPC	0D	–

6.3.3.3 The reaction term

The reaction term ($-k_r\, u(\boldsymbol{x})$) involves the evaluation of the unknown field u in the domain. Applying the same set of interpolation functions as in (6.49) to u, i.e. $u = \Sigma\beta_j f_j$, where the set β is different from the set α used in (6.49), and inverting $\boldsymbol{F}$ it is straightforward to express the potential at any point inside the domain in terms of its values at the collocation nodes. In this way, the reaction term contributes with the vector

$$\boldsymbol{b}_{REACT,i} = -\boldsymbol{k}_r \boldsymbol{u}_i\,, \quad i = 1, \ldots, N_{fn} \tag{6.60}$$

6.3.3.4 The convective term

The convective term introduces a first order derivative in space and is represented by:

$$\nabla(\mathbf{v}u) = \mathbf{v}\cdot\nabla u + u(\nabla\cdot\mathbf{v}). \tag{6.61}$$

The value of ∇u at any point inside the domain can be expressed in terms of $\mathbf{F}$ by means of:

$$\frac{\partial u(x)}{\partial x_p} = \sum_{j=1}^{N_r} \alpha_j \frac{\partial f_j(x)}{\partial x_p}. \tag{6.62}$$

And the coefficients α_j can be obtained by inversion of $\mathbf{F}$ as shown in (6.56). In the case of incompressible flow the second term in (6.61) vanishes, while the first one contributes to the discrete non-homogeneous term $\boldsymbol{b}$ in (6.57) according to:

$$\boldsymbol{b}_{CONV,i} = \sum_{k,p,j} \boldsymbol{V}_{ikp} \boldsymbol{T}_{kjp} \boldsymbol{u}_j\,, \tag{6.63}$$

where

$$\begin{aligned} V_{ikp} &= \delta_{ik} v_{kp} \\ T_{kjp} &= \frac{\partial f_{kl}}{\partial x_p} f_{lj}^{-1}, \end{aligned} \tag{6.64}$$

where the subindex p stands for the components of Cartesian coordinates (x, y, z) in the 3D case or (x', y') in 2D, while subindices i, j, k, l represent the collocation nodes in the domain.

6.3.3.5 Time integration scheme

The integration in time is based on the Finite Difference time marching scheme. A two time-level scheme has been implemented such that the time derivative is approximated in the following way:

$$\frac{\partial c}{\partial t} \approx \frac{c^{m+1} - c^m}{\delta t} \tag{6.65}$$

and the concentration c and its flux q are given by:

$$\begin{aligned} c &\approx \theta_c c^{m+1} + \left(1 - \theta_c\right) c^m \\ q &\approx \theta_q q^{m+1} + \left(1 - \theta_q\right) q^m . \end{aligned} \tag{6.66}$$

Superscripts m and $m+1$ indicate previous and present time levels, respectively. The coefficient θ_i can be adjusted from 0 to 1 yielding different schemes (Crank Nicholson, Euler implicit, Euler explicit, or any intermediate scheme). The time step δt is recalculated at every time level such that the maximum variation of concentration $\Delta c/c$ remains bounded below a certain arbitrary threshold ≈ 0.05, according to:

$$\delta t^m < \kappa \left[\frac{c}{\partial c/\partial t} \right]^{m-1} . \tag{6.67}$$

This is done in order to increase the time step when the solution is closer to its steady state, thus reducing the number of time steps and CPU requirements. This represents an important advantage when using iterative solvers, for the final system of equations.

For small problems with less than 2000 degrees of freedom, a direct solver is preferred instead of an iterative one, since the factorization of the system is calculated only once at the beginning for the initial time level, and then, further time levels involve only recalculation of the right-hand side term and matrix back-substitution operation, which are, by far, less demanding from a computational point of view than the matrix factorization. However this can be done only if the time step Δt remains fixed at every time step. As a conclusion, for small problems it is more efficient to keep Δt fixed and use a direct solver, whereas for large problems it is better to allow time step adaptivity according to expression (6.67) combined with an iterative solver.

6.3.3.6 Domain decomposition and DRM–MD

The DRM has been demonstrated to be a general and reliable procedure. However, soon it became clear that there are some problems associated with this numerical technique. The first one was related to the fact that a number of interior DRM nodes were required in order for the interpolation, see (6.49), to be more accurate. The problem associated with this was that no procedure was available for defining the optimal position of the nodes in the interior of the domain, though it was observed that this distribution significantly affects the accuracy and stability of the solution. Also, as many of the RBFs used are globally supported the matrix of the resulting system of equations is dense and frequently ill conditioned, when applied to large problems. This makes the method computationally expensive and sometimes unstable when applied to large problems. There are two ways to avoid these difficulties: by using compactly supported RBFs (CS-RBFs) or by using domain decomposition.

Domain decomposition is a technique that is commonly used in the BEM when the domain is piecewise homogeneous. After applying the numerical formulation in every subdomain, the final system of equations is obtained by means of a set of matching conditions in the interfaces between subdomains. The resulting system of equations is not dense, and the sparsity of the system increases with the number of subdomains.

A combination of domain subdivision and the DRM to avoid the domain integral was implemented by Popov and Power who called the scheme the dual reciprocity method–multi-domain (DRM–MD) approach. The initial problem solved using this formulation was the flow of a mixture of gases through a porous media [30–32]. The DRM–MD has also been applied to linear and non-linear advection-diffusion problems [33], driven cavity flow of Navier-Stokes equations [34] and of non-Newtonian fluids [35], and the flow of polymers inside mixers with complex geometries [36]. More recently the DRM–MD has been applied to a comparison of the equivalent continuum, non-homogeneous and dual porosity models for flow and transport in fractured porous media [37], 3D convection-diffusion problems [38] and flow and solute transport in 3D fractured porous media [39].

Matching conditions between two adjacent subdomains A and B must be satisfied as shown in (6.68)

$$\begin{aligned} u_A &= u_B \\ [\mathbf{q}\cdot\hat{\mathbf{n}}]_A &= -[\mathbf{q}\cdot\hat{\mathbf{n}}]_B , \end{aligned} \tag{6.68}$$

where u denotes the potential (concentration for the transport problem, or hydraulic head for the flow problem), $\mathbf{q}$ is the flux of that quantity, and $\hat{\mathbf{n}}$ is the outward normal unitary vector to the boundary of the subdomain. In this way, the formulation can deal with piecewise homogeneous material properties. Moreover, by increasing the mesh refinement it is possible to solve problems with strong variations of the material properties or the solution fields, in spite of dealing with meshes more similar to the ones employed by the finite elements method or finite volume methods.

The DRM–MD does not suffer the two main problems related to standard DRM; the systems of equations produced by DRM–MD are sparse and well conditioned, and the number and position of DRM nodes is usually not critical, since small subdomains usually require none or very few interior DRM nodes, which can be placed on equal distance in the interior of the domain. The matrices derived from the domain decomposition techniques are sparse; the higher the domain discretization the larger the number of zero entries in the matrix. This becomes a desirable advantage, since sparse matrices can be efficiently solved with many iterative techniques like Krylov based solver, conjugate gradients, etc. On the other hand the increase of discretization implies an increase in the number of degrees of freedom, this is the price to pay in order to obtain reasonable sparsity patterns. Therefore it is predictable that neither the extreme multidomain (MD) decomposition nor the classical single domain (SD) approach would offer the optimum solution in terms of computational efficiency for an arbitrary problem, but an intermediate between both.

This suggests the idea of a flexible hybrid method in which some regions of the domain can be treated with MD and some others with SD discretization. Such application of the DRM–MD technique has been suggested by Popov and Power [30, 33], but was implemented for the first time by Samardzioska and Popov [37] in 2D and by Peratta and Popov [39] in 3D. Some analysis addressing the accuracy, efficiency and stability of this approach as a function of the different discretization options is shown in [40].

6.4 Computational implementation

Bellow the main features of the developed computer code are summarized:

- The solver is based on the discrete fracture network model.
- The model for flow is based on the Darcy flow and for the transport on the advection-dispersion equation with reaction.
- The rock (3D entities), the fractures (2D entities), the fracture intersections (pipes – 1D entities) and pipe intersections (0D entities), have been implemented and coupled in a 3D code.
- The numerical approach used is the BEM with domain decomposition for the flow and boundary element–dual reciprocity method–multidomain (BE–DRM–MD) approach for the transport; as explained in the previous section.
- The computer code is implemented in such way that many subdomains with different geometries and properties can coexist in a singe model.
- The way in which the scheme is implemented offers unique flexibility to decide whether certain subdomain, 3D or 2D, would be taken as a single domain, or would be decomposed in subdomains.
- Manual or automatic fracture intersection detection.
- Automatic time step selection.
- The code is linked to the commercial package GiD for pre- and post-processing.
- The main problems encountered during the development of the method/solver were related to the different time scales of the processes in different media.

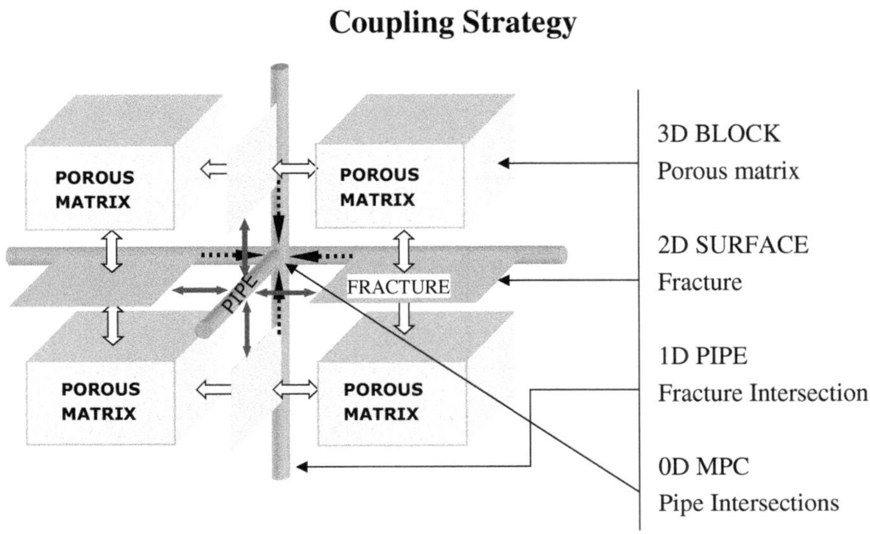

Figure 6.2: Coupling of porous matrix blocks, fractures and fracture intersections.

For example, the processes in rock and clay were of at least two orders of magnitude slower than in the fracture zones and excavation-disturbed zone (EDZ). These problems were successfully resolved by applying a semi-analytical approach.

- The code shows high accuracy and capability to integrate geometry with small details inside large-scale models.

Figure 6.2 shows the way that matrix blocks, fractures and fracture intersections interact. Fractures can be modelled as 2D or 3D entities.
The following solvers and pre-conditioners for sparse system of equations were implemented:

- direct solvers
  – LU (Gauss + Pivoting + dropping technique to keep the sparsity pattern)
- iterative solvers (sparskit) [41]
  – GMRES–FGMRES–PGMRES
  – CG
  – FOM
  – BICG
- preconditioners
  – ILUT, ILTP, ILUT(n) – Sparskit [41]
  – MC64
  – physical and geometrical scaling.

Figure 6.3 shows different possibilities available for representing 3D porous volumes. The 3D subdomains can either be discretized by volume using linear or

Shape Functions	Automatic Sub-domains Volume meshing using tetrahedrons or hexahedrons		Manual sub-domains (Geometry modeller & Surface meshing)
Quadratic Elements	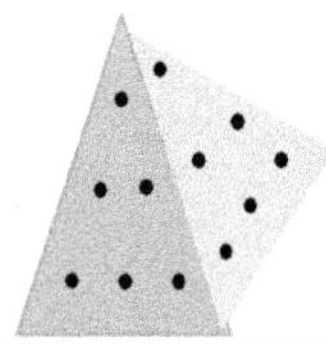	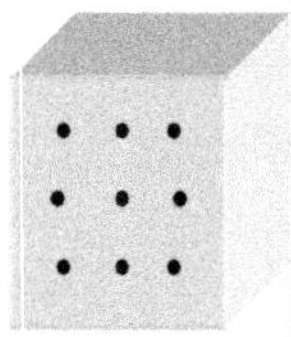	Optional internal nodes for the DRM 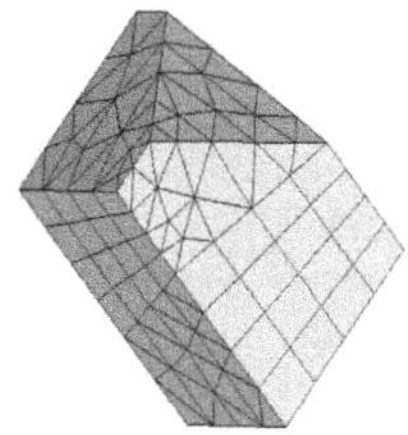
Linear Elements	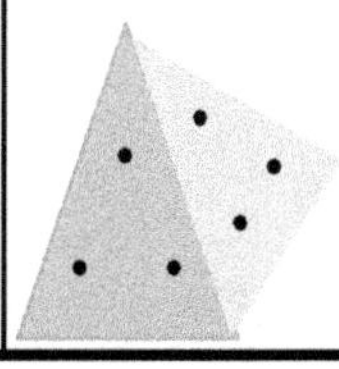	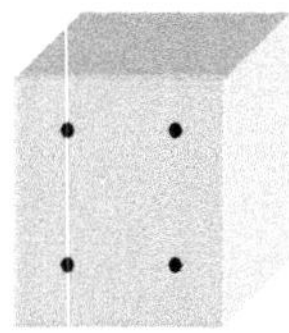	

Figure 6.3: Various possibilities for discretization of porous blocks.

quadratic tetrahedrons or hexahedrons, or can be discretized over the surface of the volume.

6.5 Results

This section contains the findings on the safety assessment of the proposed LowRiskDT approach for disposal of hazardous waste in abandoned underground mines. The safety criterion related to the waste isolating capacity of the mine repositories is evaluated taking into account the quality of the groundwater on a certain distance from the mine in the direction of the flow of the groundwater. Since no particular mine has been considered, an imaginary scenario is created where two different types of host rock media are considered: crystalline rock and limestone. All the considered parameters of the models were chosen in a conservative way or worst-case scenario, so should the findings be in favour of the approach, a sufficiently large safety margin would exist.

Both cases of mine repositories in crystalline rock and limestone were of similar geometry. The mine in crystalline rock consists of a room and a tunnel. Around the room and the tunnel EDZ was considered in the model. Eighteen fracture zones intersect the domain, of which three intersect the EDZ and serve as fast tracks for transport of contaminants. The mine in limestone consists of a room and no EDZ was considered around the repository. No fracture zones were considered for the case of mine repository in limestone.

The analysis for both mine repositories was done for two types of chemicals: dichlorvos and batteries/zinc.

Two types of analysis were performed, long-term, for periods of up to 600,000 years, and short-term, for periods of up to few thousand years.

6.5.1 Types of geological media considered

The evaluation of waste-isolating capacity in mine repositories regarding the flow and transport aspects is conducted for two different geological media, crystalline rock and limestone. In the case of crystalline rock the model consists of a large room, a tunnel, an EDZ of variable thickness around the room and the tunnel, intersecting fracture zones, and chemicals, which are embedded in compacted clay inside the room and the tunnel. In the case of limestone the model consists of a large room and chemicals, which are embedded in compacted clay inside the room.

6.5.2 The waste types considered

The waste types considered are dichlorvos and zinc (Zn).

6.5.2.1 Dichlorvos

Dichlorvos is an organophosphate insecticide with the chemical name 2,2-dichlorovinyl dimethyl phosphate. Common trade names are Astrobot, Atgard, Canogard, DDVP, and Vapona. Some of the properties of dichlorvos are as follows [42]:

- colourless liquid with a mild chemical odour, aromatic odour;
- molecular weight: 220.98;
- boiling point: 117°C at 10 mm Hg;
- solubility: 10,000 ppm at 25°C;
- vapour pressure: 1.2×10^{-2} mmHg at 20°C;
- octanol–water partition coefficient: $\log K_{ow} = 1.4$;
- chemical formula: $(CH_3O)_2(P{=}O)OCH{=}CCl_2$;

The information on the environmental fate of dichlorvos originates from studies designed to assess its use as an insecticide. Aerobic soil metabolism data showed a half-life of 0.42 days in a sandy loam soil at pH 6.2. The major metabolites were 2,2-dichloroacetic acid (DCA) (62.8% of applied dichlorvos at 48 h). Anaerobic soil metabolism in a sandy loam soil (water flooding and nitrogen atmosphere) at pH 6.8 at 25°C resulted in a half-life of 6.3 days. The major non-volatile products were DCA (accounting for up to 50.9% of applied radioactivity at day 60), 2,2-dichloroacetaldehyde (accounting for up to 12.6% of applied radioactivity at day 5), and 2,2-dichloroethanol (accounting for up to 24.7% of applied radioactivity at day 60). The potential to leach to groundwater after its application to soil is rather low, due to its rapid degradation.

As the half-life of dichlorvos is very short compared to the time-scale considered for the safety assessment, it was decided to effectively follow the transport of DCA, the main decomposition component of dichlorvos, and also the most toxic by-product. For this chemical, maximum allowable concentrations in drinking water have been imposed. For example, the US EPA has imposed a maximum contaminant level of 60 μg/L in drinking water for the average annual concentration of the sum of monochloro-, dichloro- trichloro-, monobromo-, and dibromoacetic acids [43]. The World Health Organization has a recommended level of 50 μg/L for DCA in drinking water [44] and Australia and New Zealand recommended a level of 100 μg/L [45].

The reported literature data do not mention the amount of DCA expected from dichlorvos decomposition. However, the maximum expected concentration (conservative case) of DCA would correspond to a molar ratio of DCA : dichlorvos = 1 : 1 or respective weight ratio of DCA : dichlorvos = 129 : 221 = 0.58. Thus, the maximum expected DCA concentration in mg/L would be: (maximum dichlorvos concentration, mg/L)(0.58).

Due to its high solubility in water and the presence of 2-propanol, a reasonable and also conservative sorption coefficient for DCA in all cases would be $K_d \cong 0$. The decomposition rate or the half-life for DCA is unknown and therefore following the conservative approach regarding the safety assessment, no decomposition of DCA will be considered.

6.5.2.2 Zinc

Zinc was selected as second chemical that would be considered when the waste isolating capacity in mine repositories is evaluated. Zinc would appear in the repository as a result of batteries, which would be embedded in clay in the mine repository. Zinc occurs in the +2 valence state and forms many complexes and solid phases. At pH >13 expected in the alkaline battery waste, the predominant forms of zinc are $[Zn(OH)_3]^-$ and $[Zn(OH)_4]^{2-}$, in equilibrium with $Zn(OH)_{2(s)}$. If significant concentrations of chloride anions (halite rock, sea water) or carbonate anions (carbonate rocks) are present, additional soluble species, such as $ZnCO_{3(aq)}$ and $ZnCl^+$ may be formed. In Leclanche cells, the solid phases $Zn(NH_3)_2Cl_{2(s)}$ and $ZnCl_2.4Zn(OH)_{2(s)}$ have been reported.

6.5.3 Case of mine and tunnel in crystalline rock

The first case is a mine repository in crystalline rock, which includes a room and a tunnel, where both the room and the tunnel are filled with chemical waste embedded in clay.

6.5.3.1 Geometry definition

Figures 6.4 and 6.5 show the side view and the top view of the considered domain, respectively, including the geometry of the mine and the tunnel. The dimension of the room is 100 m × 50 m × 50 m. The room is filled with hazardous waste embedded in compacted clay. The length of the tunnel is 150 m and the

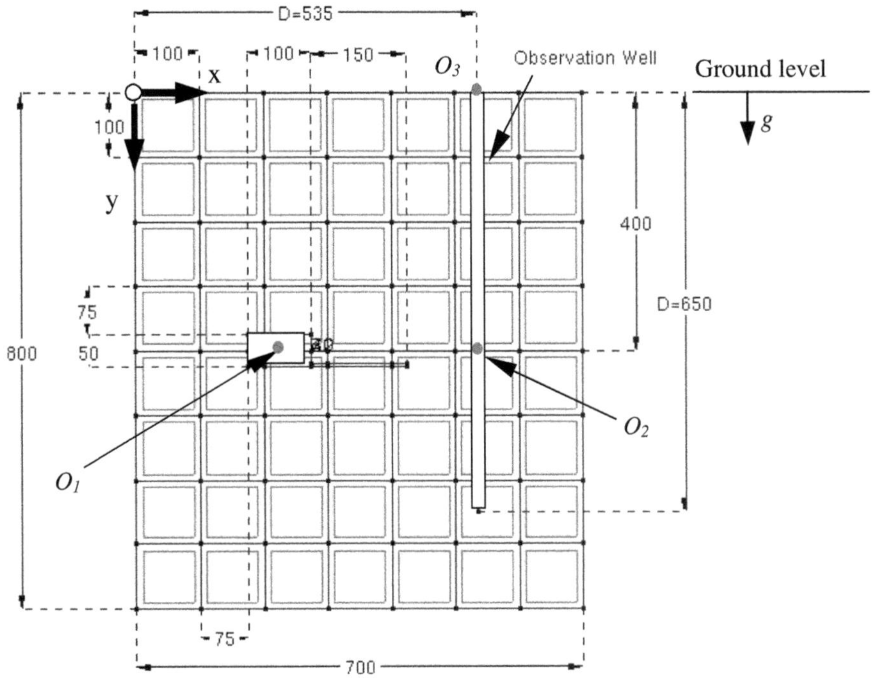

Figure 6.4: Side view of the considered domain showing the geometry of the mine and tunnel and the respective position of the observation well.

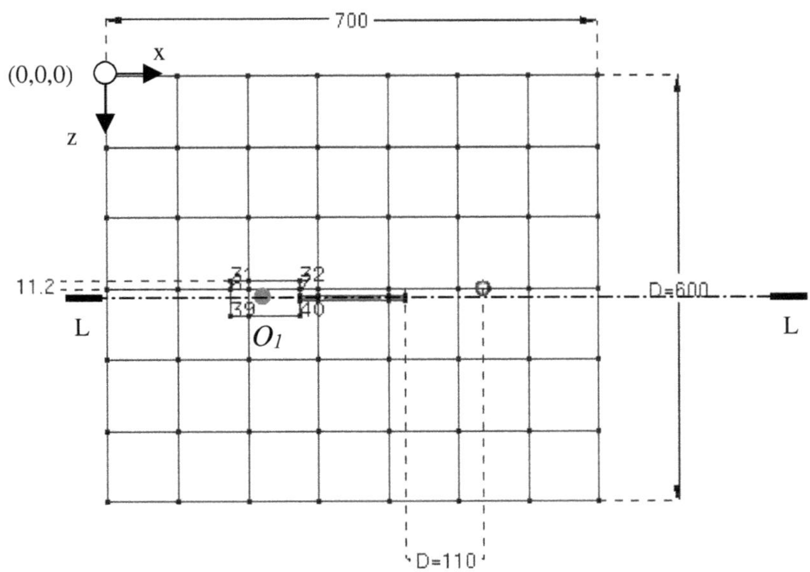

Figure 6.5: Top view of the considered domain showing the geometry of the mine and tunnel and the respective position of the observation well.

Table 6.3: Definition of observation points.

Point	Position	Absolute coordinates (x, y, z)
O_1	Geometrical midpoint of the room	225, –400, 300
O_2	In the well, at the room level	535, –400, 300
O_3	In the well, at ground level	535, 0, 300

cross section is 5 m × 5 m. Around the room and the tunnel there is an EDZ with variable thickness, which is not shown in the figures. The width of the EDZ is 3 m around the room and 1 m around the tunnel. The vertical and horizontal lines in the figures represent fracture zones with a width of 1 m. It can be seen that there are 18 fracture zones in the model. The fracture zones are perpendicular to each other, separating in this way large porous rock blocks, being each one of them surrounded by six fractures. In this case perpendicular flat fracture zones were selected for simplicity reasons, when defining the fracture network, though the model and the computer code can cope with any shape and distribution of the fracture network. For each porous block the average hydraulic properties are considered.

In Fig. 6.4 it can be seen that one fracture perpendicular to x-axis and one fracture perpendicular to y-axis intersect the EDZ of the room, while two fractures perpendicular to x-axis intersect the tunnel. In Fig. 6.5 it can be seen that one fracture perpendicular to z-axis intersects the EDZ of the room. This is considered to be a conservative case since three fracture zones intersect the EDZ of the room and two intersect the EDZ of the tunnel, which speeds up the transport of the chemicals, which would leak out of the repository. Figures 6.4 and 6.5 and Table 6.3 show the position of the observation well and observation points. The observation well and observation points are used in order to simulate the appearance of the chemicals in a certain place in the domain. It has been selected that the well is inside the fracture that intersects the EDZ of the room, as the fastest transport would exist in this fracture. Also, for the same fracture that intersects the room, which is perpendicular to the z-axis and is shown in Fig. 6.5, concentration change in time is followed on the top of the domain, or, the surface of the rock massif.

6.5.3.2 Model discretization

In Fig. 6.6 the discretization of the domain into rock blocks, fractures and fracture intersections is shown. The position of the room and the tunnel in the model is shown in Fig. 6.7. The full model is divided into two domains, near field, represented with green colour in Fig. 6.6, and far field. The rock blocks in the near field are discretized by volume and the rock blocks in the far field are discretized on the boundary only. This gives possibility to obtain more information about the processes in the near field, to take into account the processes in the far field and

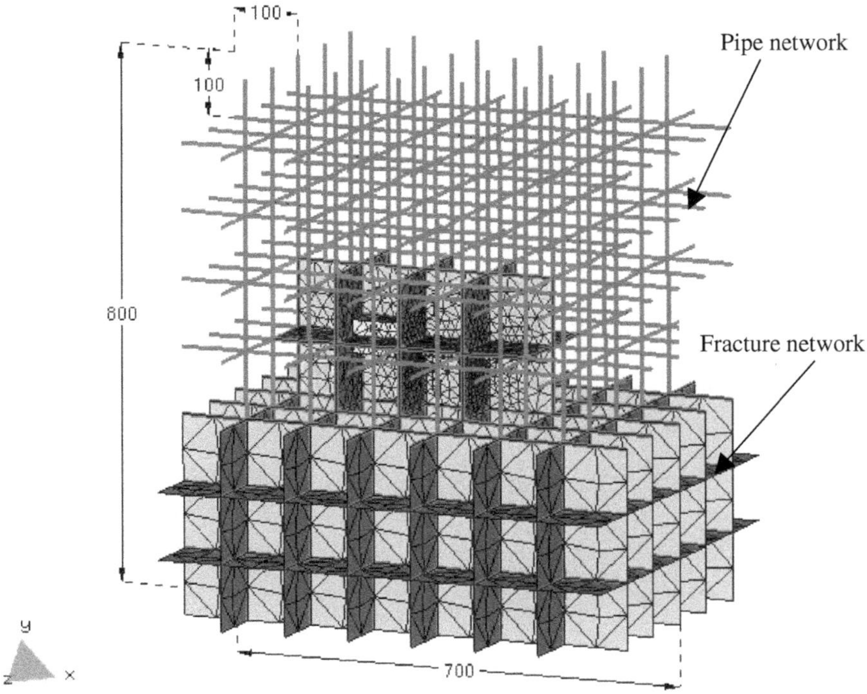

Figure 6.6: Discretization of the domain into rock blocks, fracture zones and fracture intersections (pipes).

still to keep the size of the system of equations within desired limits. Table 6.4 shows the number of different entities included in the model (3D porous blocks, fractures, fracture intersections).

6.5.3.3 Parameter estimation

The hydraulic and transport properties of the rock, fracture zones, clay and EDZ are shown in Table 6.5. The hydraulic conductivities were chosen of order of magnitude that can be considered to be conservative. In other words, considering that the room is in crystalline rock, it is not expected that the actual hydraulic conductivities would be higher than the used ones.

6.5.3.4 Boundary and initial conditions

The boundary conditions for flow are shown in Fig. 6.8. On the top surface atmospheric pressure and on the bottom surface impermeable boundary conditions are imposed. One of the vertical surfaces has got 5% overpressure in respect to the hydrostatic pressure, while the other three vertical surfaces are with hydrostatic pressure. The hydrostatic pressure is not shown in Fig. 6.8. Saturated flow is assumed.

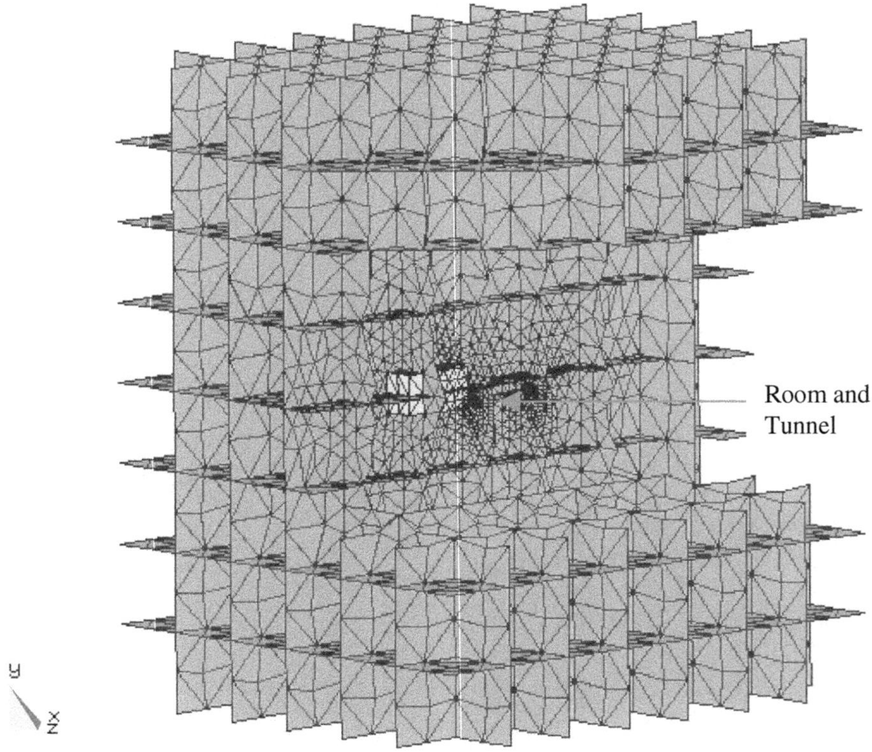

Figure 6.7: Position of the room and the tunnel in the model.

Table 6.4: Number of geometrical structures included in the model.

Number of freedom nodes	188352
Number of geometrical nodes	7526
Number of DRM nodes	14212
Number of 3D blocks	340
Number of 3D subdomains	40290
Number of 2D subdomains	10636
Number of 1D subdomains	1788
Number of pipe intersections	1468

The initial conditions for transport are given by defining the amount of chemical that is stored in the room and tunnel. In other words, the concentration of the chemical in the repository will decrease in time as it will leak from the repository in the surrounding strata. It is assumed that at time $t = 0$ there is no hazardous chemical present outside the repository.

Table 6.5: Hydraulic and transport properties of different entities, which form the geometry of the case study.

Parameter	Clay buffer	Rock–porous matrix	Fracture zones	EDZ
K (m/s)	1×10^{-10}	1×10^{-9}	1×10^{-7}	1×10^{-7}
D (m^2/s)	1×10^{-10}	1×10^{-9}	1×10^{-7}	1×10^{-7}
R	1	1	1	1
k_r (1/s)	0	0	0	0
Width (m)	–	–	1	3

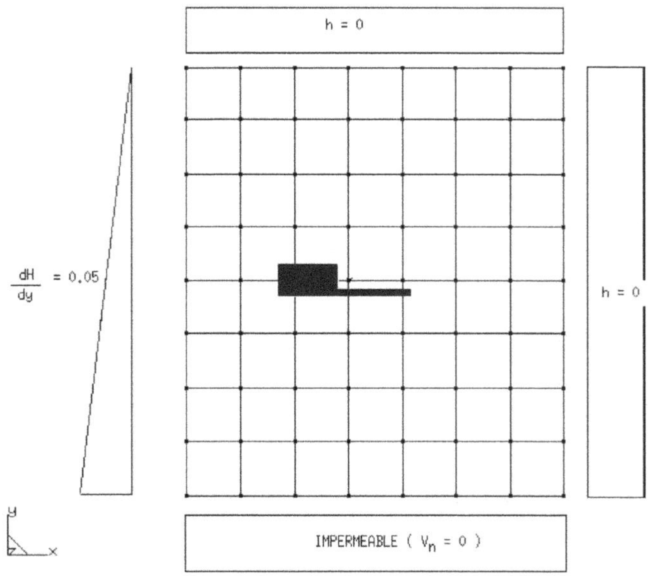

Figure 6.8: Flow conditions for the crystalline case.

6.5.3.5 Results for flow

The results for the flow in the case of crystalline rock are shown in Fig. 6.9. Figure 6.10 shows a close-up view of the hydraulic head and velocity field near the room. The hydraulic head is represented as a density plot over the fracture network, whereas the velocity field is represented as a vector plot. Typical values of the velocity in the rock matrix are two orders of magnitude lower than in the fracture network, which is due to the difference in hydraulic conductivities in rock and fractures. The maximum gradient of hydraulic head over the whole domain is $\nabla h = 0.05$. The influence of the EDZ can be seen in Fig. 6.10. The flow is directed towards the EDZ on the inflow part of the EDZ/room, in the

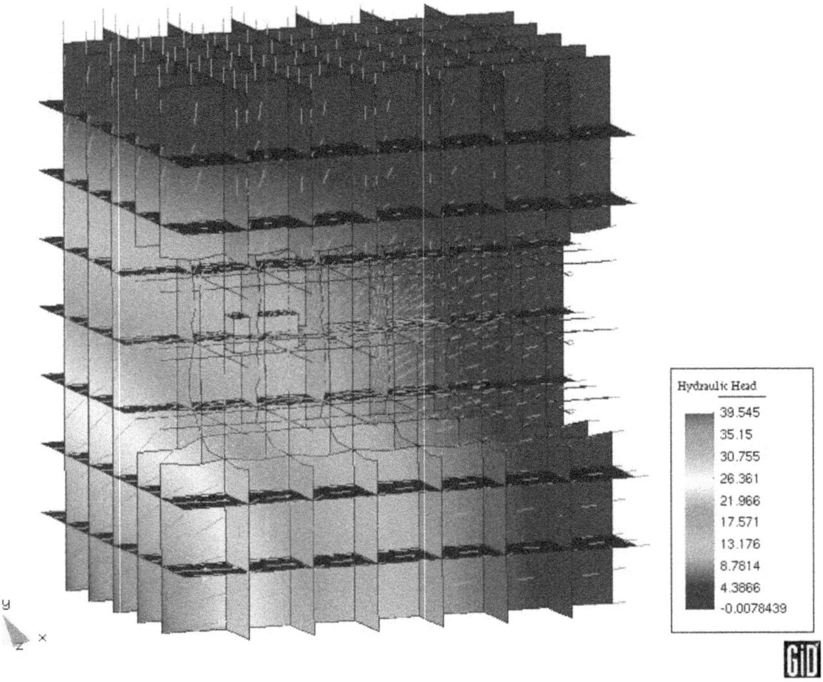

Figure 6.9: Density plot of hydraulic head and vector plot of Darcy velocity in the fracture network.

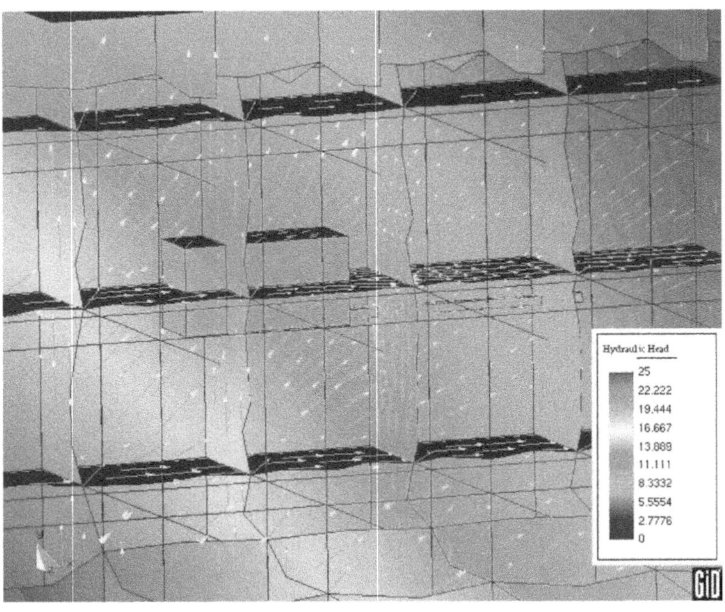

Figure 6.10: Detail of the velocity field and hydraulic head close to the near field.

figure bottom-left side of the room, and away from the EDZ/room on the outflow part, in the figure top-right side of the room. This is a consequence of the higher hydraulic conductivity of the EDZ in respect of the surrounding rock, making the EDZ take larger part of the flow in the vicinity of the room.

The effect of the EDZ is more evident in Fig. 6.11, which shows only the velocity field.

6.5.4 Case of disposal of dichlorvos in mine repository in crystalline rock

The first case considered is for the chemical dichlorvos disposed of in underground mine. In the following sections the safety aspects of a hypothetic abandoned underground mine in which dichlorvos is disposed of in clay will be considered.

6.5.4.1 Modelling conditions for dichlorvos

The transport is calculated for the DCA, product of decomposition of dichlorvos. The distribution coefficient K_d is taken to be 0, therefore, the retardation factor R is 1. The worst-case scenario is considered, when there is no decay of DCA.

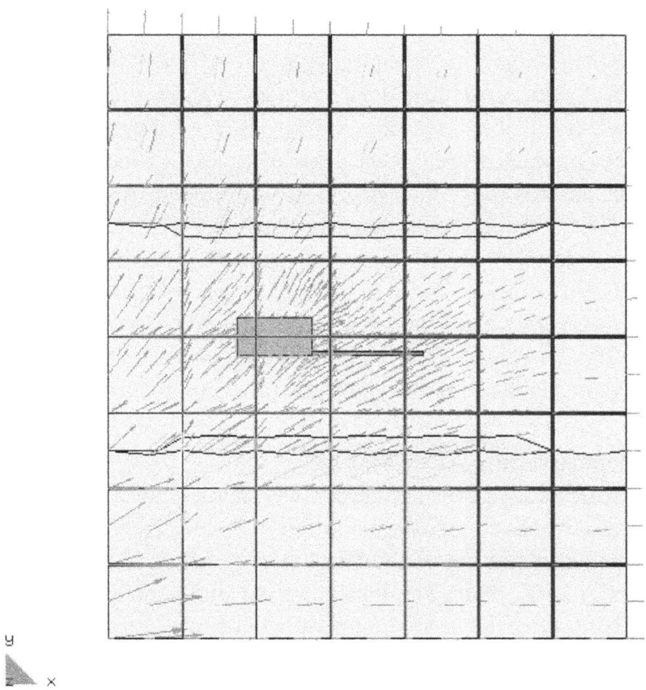

Figure 6.11: Vector plot of Darcy velocity in the fracture network.

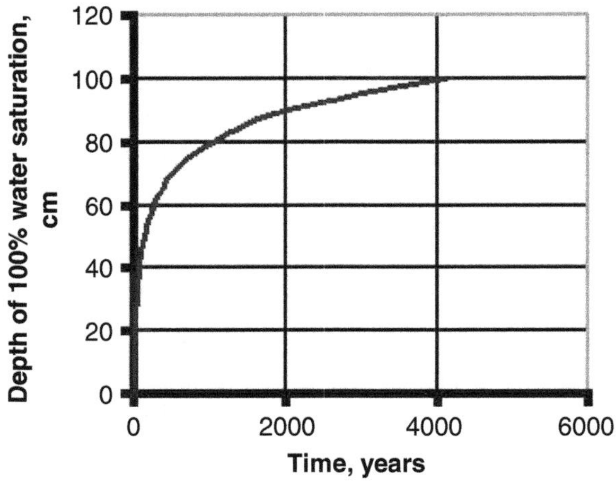

Figure 6.12: Movement of the wetting front assuming only diffusion with $D = 10^{-9}$ m^2/s.

The simulation is done by considering that there is certain amount of solidified dichlorvos embedded in clay in the repository at time t = 0. Only DCA is followed, as dichlorvos, because of the very short half-life, does not travel far and very soon after deposition, in maters of months, completely decays. Note that the conservative approach has been followed in respect to saturation of clay, as it was considered that the clay is saturated at t = 0. The LowRiskDT document D2.2 [46] shows that for the Friedland Ton clay with density at water saturation of 1900 kg/m^3, practically important wetting of the volume of clay-embedded waste will not commence until 4000 years after application, providing that the waste mass in the big room in this case is surrounded by a 100 cm 'liner' of Friedland Ton with the assumed density, see Fig. 6.12. This result is obtained without taking into account the impact of pressure-induced wetting. The initial concentration of dichlorvos in the repository is assumed to be 10,000 ppm. As part of the conservative analysis it is assumed that dichlorvos completely transforms into DCA. Therefore, shortly after the start of the simulation, the concentration of DCA inside the repository is equal to 10,000 ppm.

6.5.4.2 Transport results for dichlorvos

As a test of the accuracy of the simulation we use the plot of the concentration in the 'pipe'/fracture intersection, shown in Fig. 6.13 as a parallel bright line above the tunnel. The concentration of DCA in Fig. 6.13 has been normalized, where 1 corresponds to 10,000 ppm. The concentration plot, Fig. 6.14, starts after the room and finishes after the second perpendicular fracture that intersects the tunnel. It is evident that there is a jump in the concentration related to the point where the second perpendicular fracture connects the pipe with the tunnel. Thought there were many tests of the 3D solver towards various 1D analytical

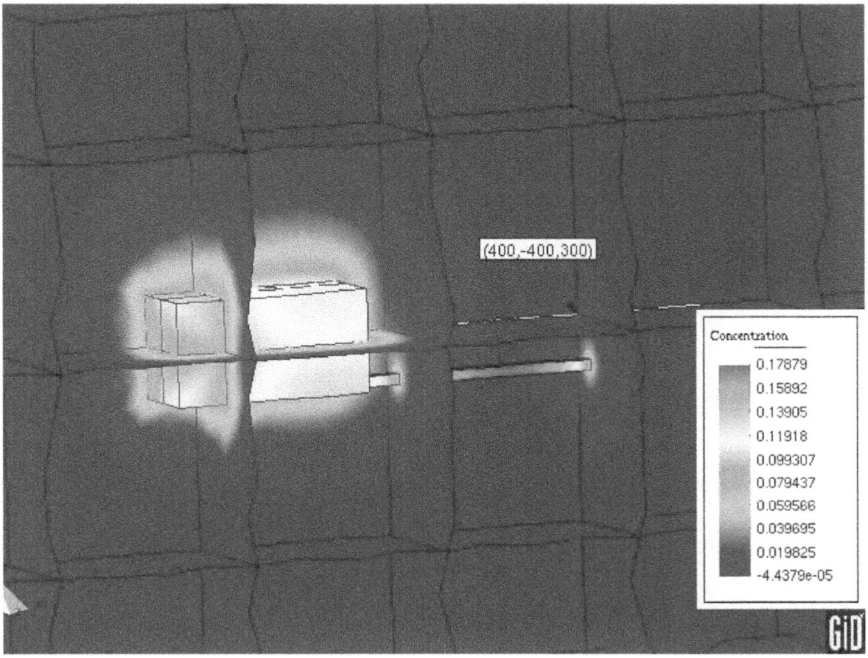

Figure 6.13: Close-view of transport of DCA after 1000 years in the vicinity of the repository.

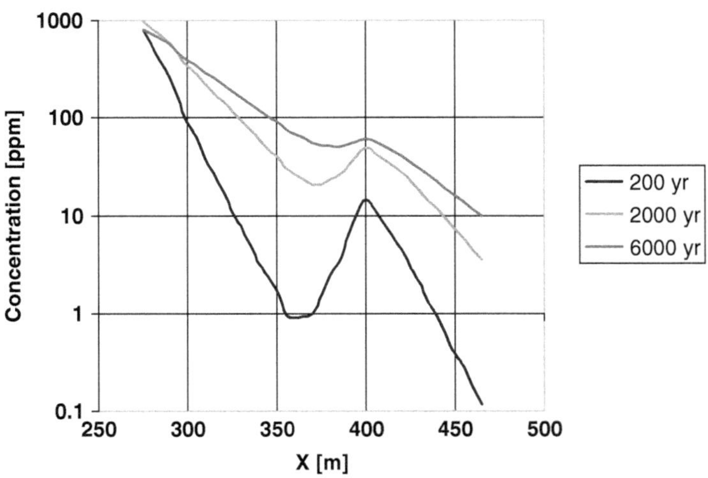

Figure 6.14: The effect of the tunnel in the crystalline case, where a vertical fracture is crossing the tunnel at $x = 400$ m and affecting the transport in the fracture intersection under observation.

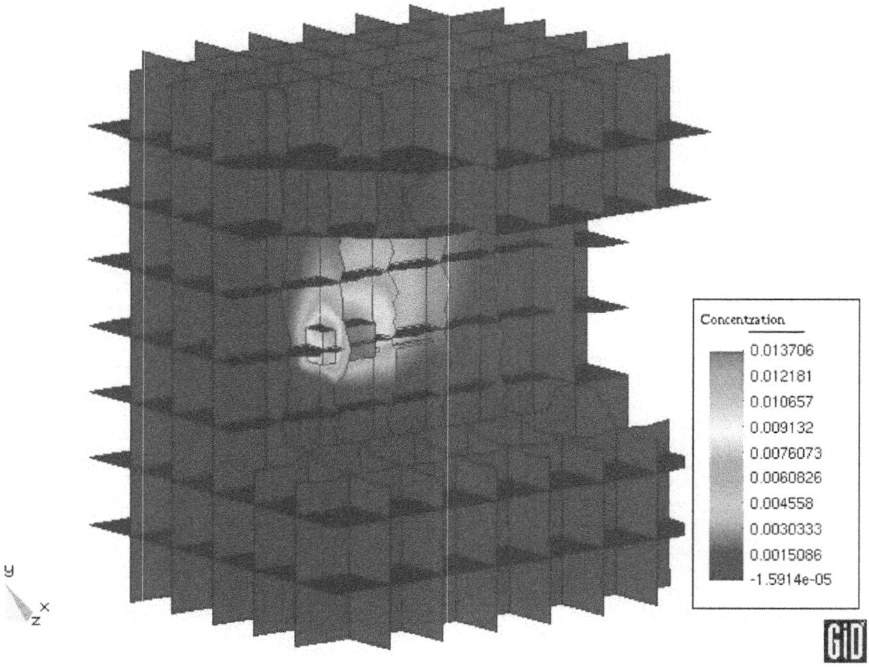

Figure 6.15: Normalized concentration distribution of DCA in the considered domain after 127,000 years.

solutions that verified the accuracy, this example confirms the accuracy of the model in real 3D conditions.

Figures 6.15 and 6.16 show the character of the transport. Since the hazardous waste is embedded in clay of low hydraulic conductivity, the chemical is slowly released in time. This is a case of contaminant source which changes the magnitude in time as the concentration inside the repository drops down. It can be seen that by $t = 100{,}000$ years the maximum concentration in the rock is still close to the repository since the concentration gradient inside the repository is still high enough to induce significant flux of contaminant. At $t = 300{,}000$ years the outflux from the repository has decreased significantly, so the highest concentration in the rock is not adjacent to the repository any more, and is due to contaminant, which escaped the repository in the past. Both effects in the transport of the chemicals are evident, the advection and the dispersion. The timescale of the whole process of release of the chemical from the repository in this case is of the order of magnitude of 600,000 years, and it is referred to this time-scale as a long-term analysis. The short-term analysis is of the order of few thousand years. One example of short-term analysis is shown in Fig. 6.13, situation shown at $t = 1000$ years.

Figure 6.17 shows the results for long-term analysis of DCA transport through a fracture that intersects the EDZ of the repository and appears at the ground level.

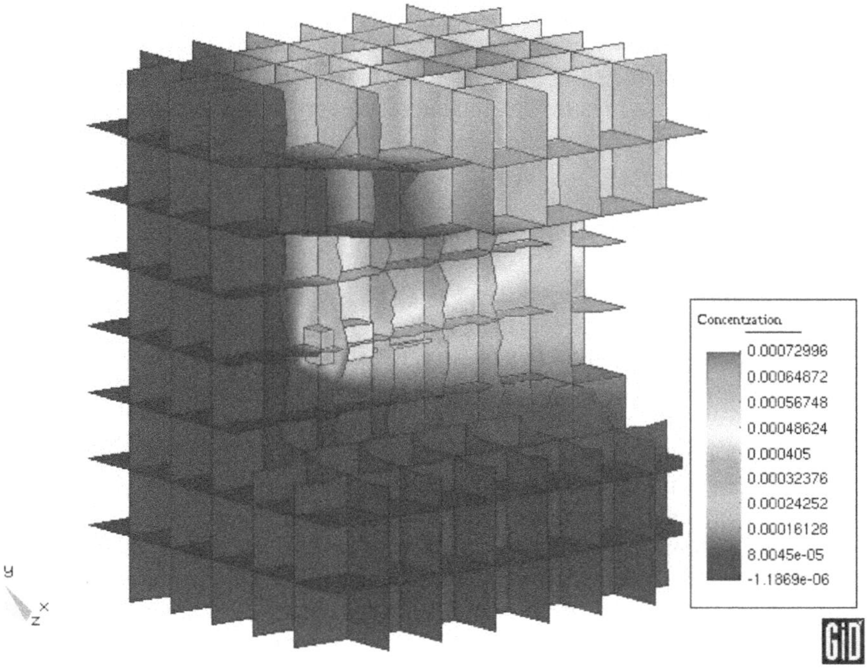

Figure 6.16: Normalized concentration distribution of DCA in the considered domain after 317,000 years.

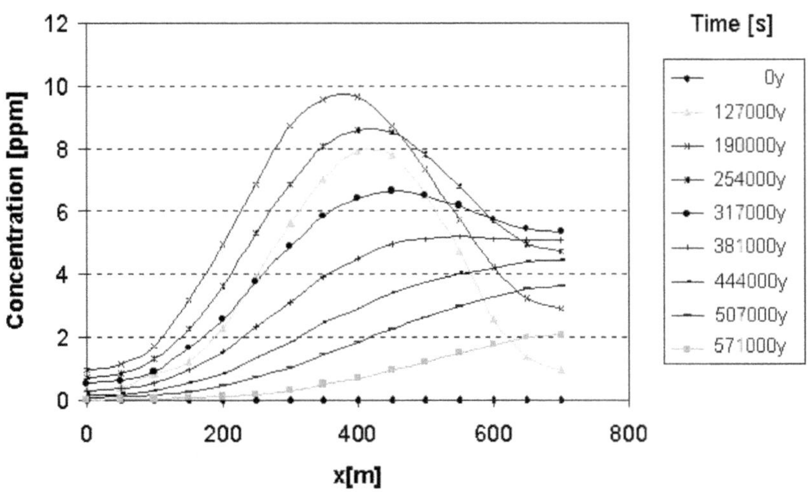

Figure 6.17: Long-term analysis of DCA concentration variation on the ground level in the fracture closest to the LL line (see Fig. 6.5).

The fracture can be seen in Fig. 6.5 and is the closest one to the LL line. The analysis has captured the moment of highest concentration at the ground level, approximately $t = 200{,}000$ years. The value for the highest concentration is close to 10 ppm. However, these results are obtained for a very conservative case and in a real case it is highly unlikely that the concentrations would reach such level, because of the following:

1. Not all of the dichlorvos would transform/decay into DCA.
2. There would be decay of DCA and it may be that no DCA ever reach the surface. However, no data on decay of DCA is available.
3. Fully saturated flow was considered.
4. The flow velocity vector, see Fig. 6.11, has got a significant upward (towards the surface) component, which in reality may not be the case.

The other aspect that must be mentioned is that any analysis longer than few thousand years is unreliable since by that time tectonic as well as glacial processes may completely change the situation.

Figure 6.18 shows concentration variation in time in the middle of the room, point O_1, and in the observation well, point O_2, see Fig. 6.4. It can be seen that by 300,000 years very little of the contaminant left inside the repository. The maximum concentration inside the well in point O_2 is estimated at $t \approx 200{,}000$ years. The maximum concentration in point O_2 is approximately 3 ppm, which is lower than the maximum concentration found in the point O_3 in the well at the surface, see concentration of DCA in Fig. 6.17 for $x = 535$ m. The highest concentration in O_3 is above 7 ppm. This effect is due to the vertical component of velocity

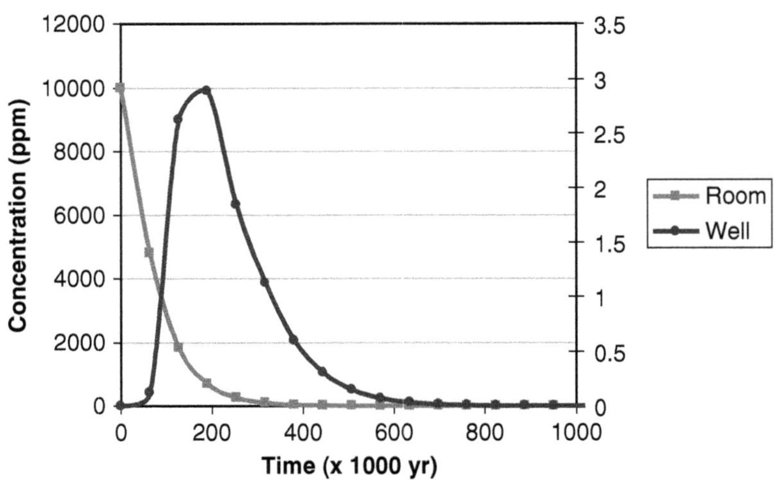

Figure 6.18: DCA concentration variation in time in the middle of the room, point O_1 in Fig. 6.4, shown on the left axis, and in the well, point O_2 in Fig. 6.4, shown on the right axis.

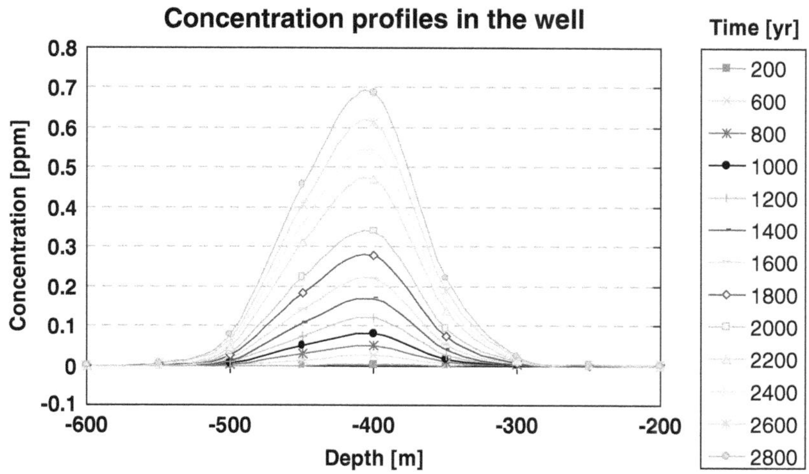

Figure 6.19: Short-term DCA concentration variation in time in the well.

vector, see Figs 6.11 and 6.16. Figure 6.17 shows that the concentration distribution at the surface changes in time due to the advection dispersion processes. The dispersion process makes the peak wider in time, while the advection process moves the location of the peak in the direction of the groundwater flow.

Figure 6.19 shows the short-term analysis of the concentration of DCA in the observation well. It is evident that the concentration in the point O_2 is higher than the concentration in the well at the surface, in the first few thousand years. The DCA concentration in point O_3 is not shown in Fig. 6.19 as at $y = 0$ m DCA is still not present after 3000 years. The concentration of DCA is higher in O_2 than in O_3 due to the smaller distance between the point O_2 and the tunnel and the room and also because two fractures intersect at O_2, one vertical and one horizontal. These two fractures also intersect the EDZ, see Figs 6.4 and 6.5, which increases the transport rate. DCA reaches O_3 later than it reaches O_2, however, the maximum concentration of DCA in O_3 at later stages exceeds the one in O_2 for the reasons explained above.

It is worth noting that the situation in O_2 would be different than what is shown in Fig. 6.19 due to several factors:

1. It was considered that the clay is saturated at $t = 0$ years. As mentioned above, in the case of the Friedland Ton clay with density at water saturation of 1900 kg/m^3, practically important wetting of the volume of clay-embedded waste will not commence until 4000 years after application, see Fig. 6.12, providing that the waste mass in the big room in this case is surrounded by a 100 cm 'liner' of Friedland Ton with the assumed density. Therefore, at time $t = 3000$ years the chemical would have not left the repository at all, or very little would have leaked.
2. Not all of the dichlorvos would transform/decay into DCA.

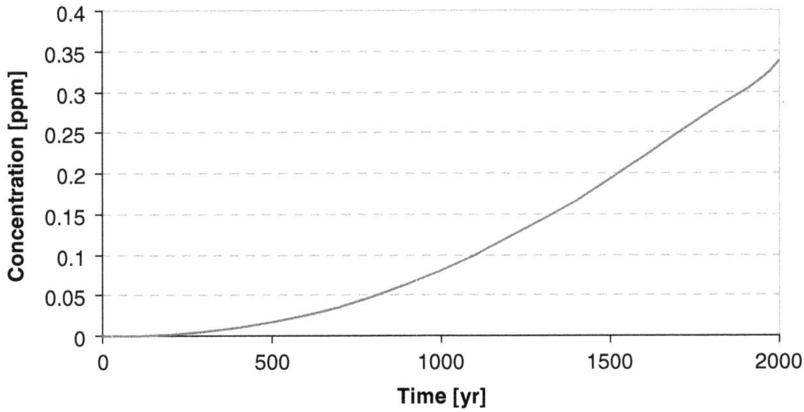

Figure 6.20: DCA concentration variation in time in O_2 during the first 2000 years.

3. There would be decay of DCA.
4. Fully saturated flow was considered, which would not be the case in general.

Even under such conservative/unrealistic assumptions, the concentration of DCA in O_2 would not exceed 1 ppm in 3000 years. The results in Fig. 6.19 show that after 600 years the concentration in O_2 will not exceed 0.05 ppm, which is equivalent to 50 μg/L, and is within the recommended highest concentration allowable by the World Health Organization of 50 μg/L for DCA in drinking water [44].

Figure 6.20 shows the DCA concentration variation in O_2 in the first 2000 years.

6.5.5 Case of disposal of zinc in mine repository in crystalline rock

Second analysis for mine repository in crystalline rock was performed for the case of waste containing batteries. In the following sections the safety aspect of disposal of batteries in abandoned underground mines will be discussed.

6.5.5.1 Modelling conditions for zinc

The geometry of the mine tunnel and considered domain and all the parameters in the model remained the same except the ones mentioned below.

It was considered that the batteries are mixed with Friedland Ton clay in ratio 50 : 50 by volume. It was assumed that at $t = 0$ years the clay is fully saturated and the batteries are decayed in a way that the zinc can travel through the clay and into the surrounding rock. It was further assumed that 25% of the batteries mass is due to zinc. In the case of zinc it was taken that K_d is 0.02 m^3/kg and using the following equation

$$R = 1 + \frac{\rho}{m} K_d$$

it is estimated that the retardation factor is approximately $R = 121$.

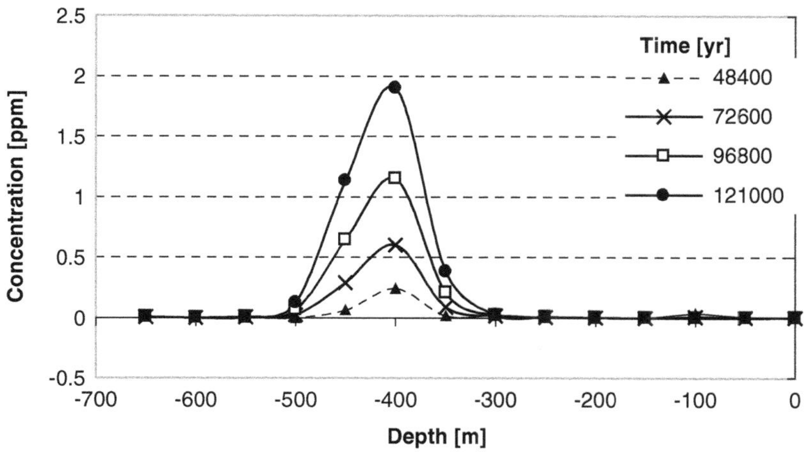

Figure 6.21: Zinc concentration variation in time in the well.

6.5.5.2 Transport results for zinc

The results for leakage and transport of zinc are shown in Fig. 6.21. It can be seen that in this case the 'short-term' should refer to the period of 100,000 years. This is due to the retardation of the zinc during the transport through the surrounding rock, an effect which did not exist in the case of DCA. The same velocity field as in the case of dichlorvos/DCA is obtained for zinc, shown in Figs 6.9–6.11. Similar propagation pattern to the one shown in Figs 6.15–6.20 for dichlorvos/DCA is obtained for zinc, with the difference of slower propagation rate.

Results in Fig. 6.21 show that in the period of 120,000 years the concentration of zinc in the observation well will not exceed 2 ppm. However, these results are obtained for a very conservative case and in a real case it is very unlikely that the concentrations would reach such level, because of the following factors:

1. In the analysis it was assumed that the clay is saturated at $t = 0$ years.
2. Fully saturated flow is considered.
3. It is considered that the batteries have decayed at $t = 0$ in such way that zinc is free to travel through the clay and surrounding rock.
4. It is not considered that the rock would absorb and immobilize part of the zinc, which would reduce the amount of zinc available for transport. The influence of the retardation factor R in the equations is to slow down the transport, it does not immobilize the zinc.

6.5.6 Case of mine in limestone

The second case is a mine repository in limestone, which includes a room filled with batteries embedded in clay.

Table 6.6: Hydraulic and transport properties of different entities, which form the geometry of the case study.

Parameter	Clay buffer	Rock
K (m/s)	1×10^{-10}	1×10^{-6}
D (m^2/s)	1×10^{-10}	1×10^{-6}
R	1	1

6.5.6.1 Geometry definition

The geometry of the mine and domain is the same one that is shown in Figs 6.4 and 6.5, with the difference that in this case the following was excluded from the model: fractures, tunnel and EDZ. The observation well remains in the same place, see Figs 6.4 and 6.5.

6.5.6.2 Parameter estimation

The hydraulic and transport properties of the rock and clay are shown in Table 6.6. The hydraulic conductivities were chosen of order of magnitude that can be considered to be conservative. In other words, considering that the room is in limestone, it is not expected in reality that larger values for hydraulic conductivities would exist, than the ones used in the examples.

6.5.6.3 Boundary and initial conditions

The boundary and the initial conditions are the same as in the case of crystalline rock given in Section 3.2.4.

6.5.6.4 Results for flow

The results for flow are given in Figs 6.22–6.26. Since there are no fractures in this model, the hydraulic head and the velocity field are shown in an arbitrary plane, see Fig. 6.22. Figure 6.23 shows part of the mesh used in the model, as well as the plane used to show the results. Figures 6.24 and 6.25 show the side view and the top view of the velocity field with the hydraulic head. Figure 6.26 shows the velocity field and hydraulic head in vicinity of the repository. In the figure the characteristic velocities of the model are shown in some points. It can be seen that the velocities in the limestone are of the order of 1 m/year, while inside the repository the velocities are of the order of 0.1 mm/year. This shows that the difference in the velocities is of four orders of magnitude, which is in agreement with the difference in hydraulic conductivities in the clay and limestone. The transport in the clay is predominantly by diffusion, while in the limestone it is combined, advection and dispersion.

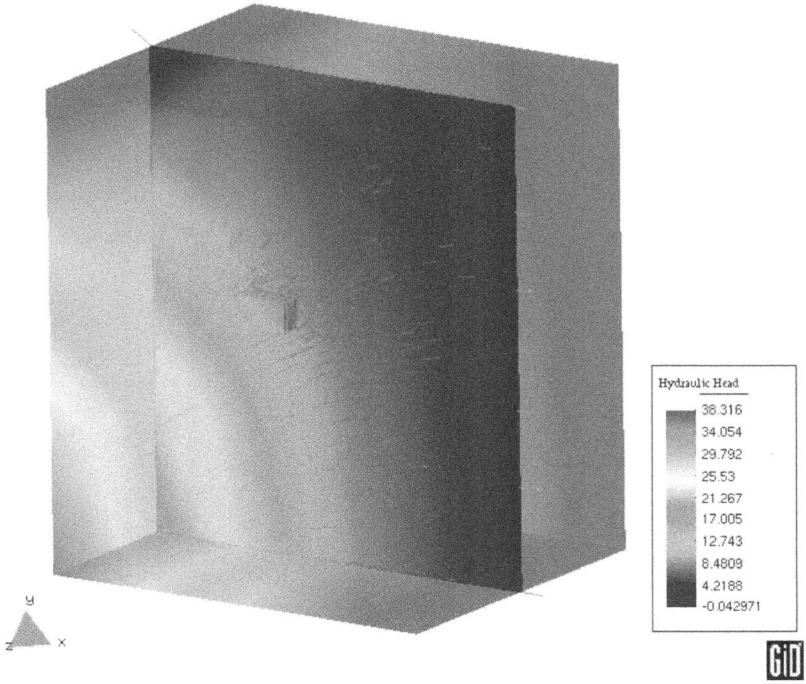

Figure 6.22: Hydraulic head and velocity field for the case of mine repository filled with clay in limestone.

In this case since the EDZ is not included in the model, one can observe different velocity field than the one for the case of repository in crystalline rock. While in crystalline rock the EDZ makes the streamlines converge towards the repository and diverge from the repository, because of the higher permeability, see Fig. 6.10, in the case of repository in limestone, the absence of the EDZ makes the streamlines diverge towards the repository and converge from the repository, see Figs 6.25 and 6.26.

6.5.7 Case of disposal of dichlorvos in mine repository in limestone

In the following sections the safety aspects of a hypothetic abandoned underground mine in which dichlorvos is disposed of in clay will be considered.

6.5.7.1 Modelling conditions for dichlorvos

The modelling conditions for dichlorvos were the same ones that were used in the case of mine repository in crystalline rock and are described in Section 6.5.4.1.

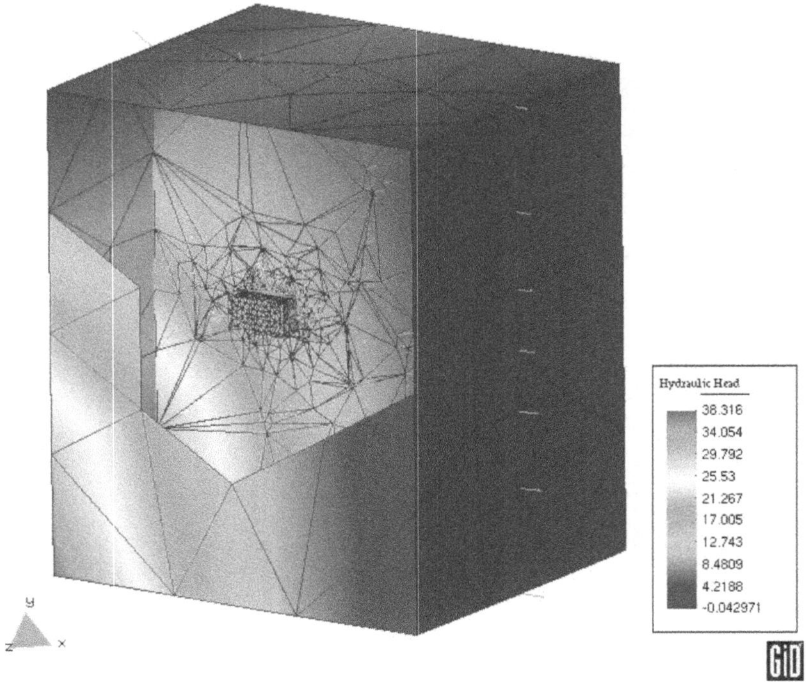

Figure 6.23: Hydraulic head, velocity field and part of the model mesh used for the case of mine repository in limestone.

6.5.7.2 Transport results for dichlorvos

Figures 6.27 to 6.31 show the results for leakage of DCA from the repository and its transport through limestone. All the results are for short-term analysis, up to 2000 years. The process of leakage and transport is similar to the one in crystalline rock where the chemical is slowly released mainly by diffusion, because of the low hydraulic conductivity of the Friedland Ton clay, and after that it is relatively rapidly transported through the limestone due to both, advection and dispersion. In this sense the process looks like a quasi steady-state as the distribution of the DCA in the space looks similar in respect to the highest concentration in the domain, what changes are the concentrations which decrease due to the decrease of DCA inside the repository, which in turn reduces the outflux of DCA.

There is difference in the processes of transport of DCA once it leaves the repository, depending on whether the mine is in crystalline rock or limestone. In crystalline rock the main transport is conducted through fractures and fracture zones, since there the hydraulic conductivity is much higher than in the rock. The crystalline rock slows down the transport by absorbing the chemical, which penetrates the rock mainly by diffusion. In the case of repository in crystalline rock the transport will be mainly defined by the characteristics and

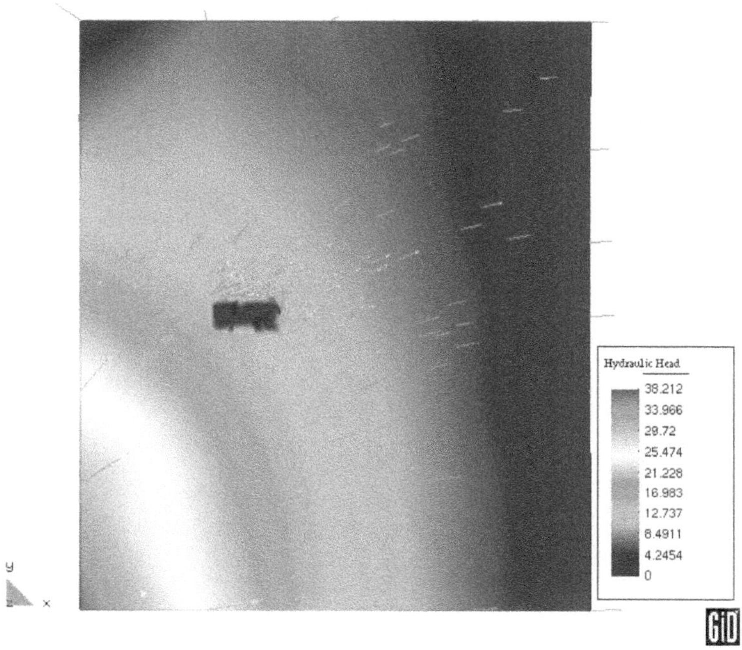

Figure 6.24: Side view of the hydraulic head and velocity field for the case of mine repository in limestone.

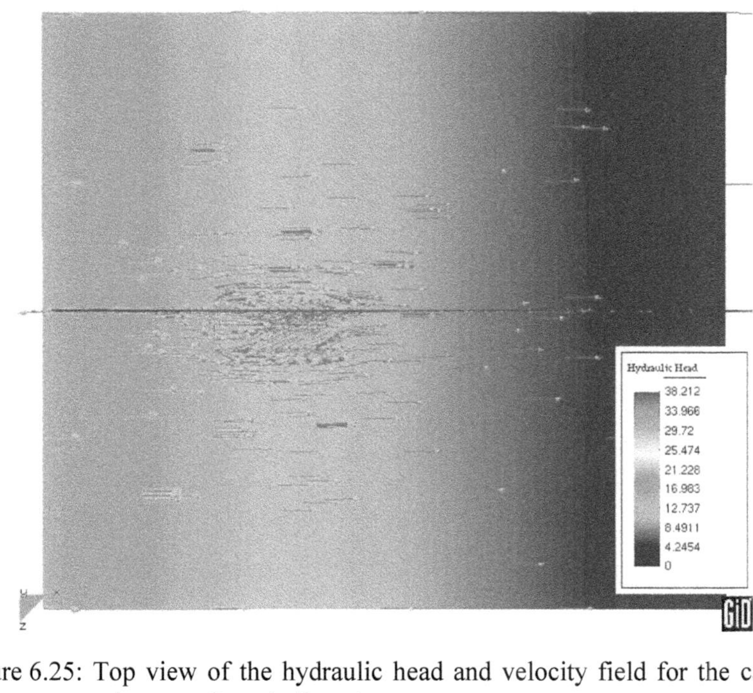

Figure 6.25: Top view of the hydraulic head and velocity field for the case of mine repository in limestone.

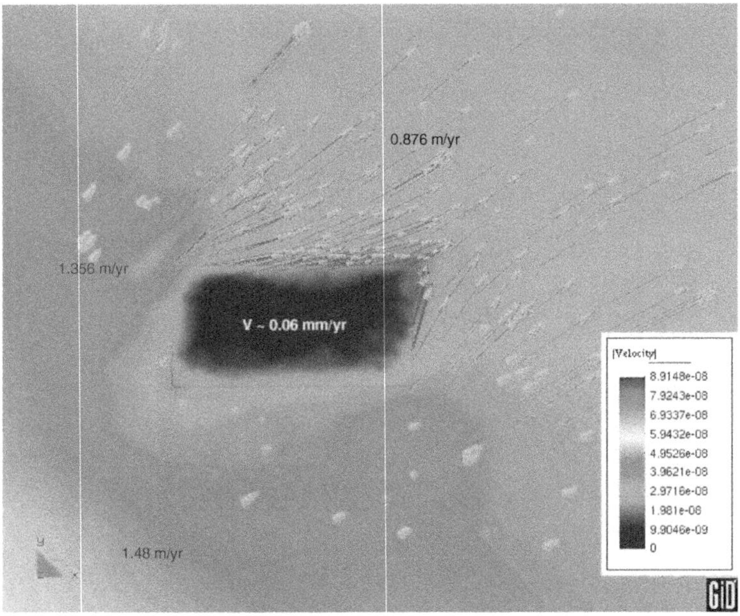

Figure 6.26: Side view of the hydraulic head and velocity field near the mine repository in limestone.

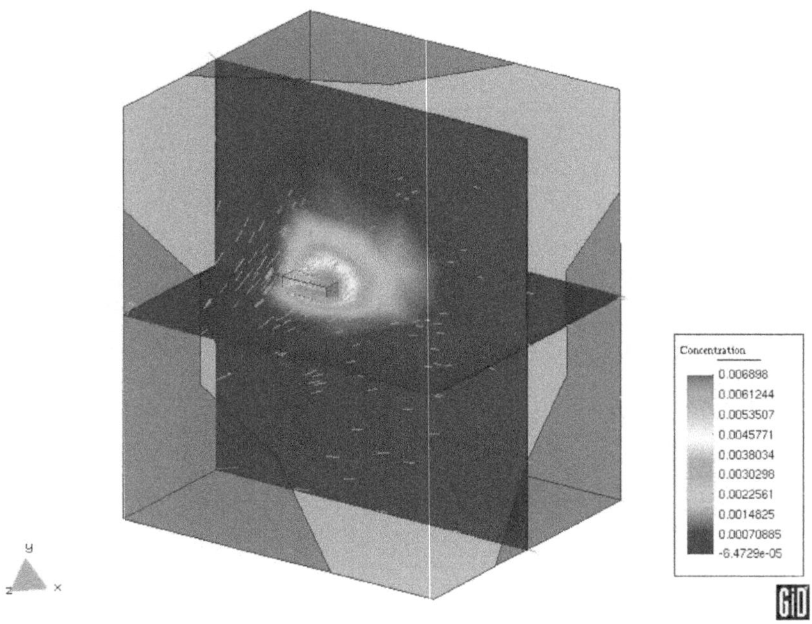

Figure 6.27: Normalized concentration distribution of DCA in limestone around mine repository after 200 years.

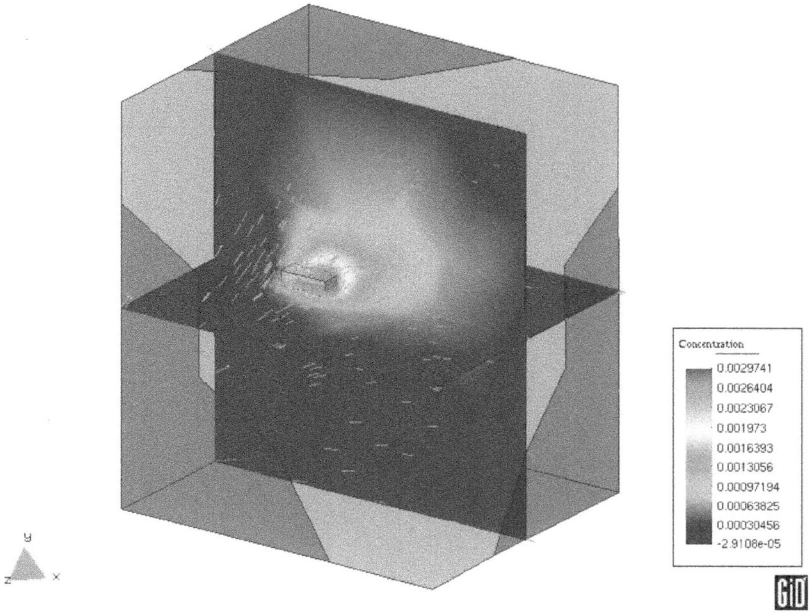

Figure 6.28: Normalized concentration distribution of DCA in limestone around mine repository after 400 years.

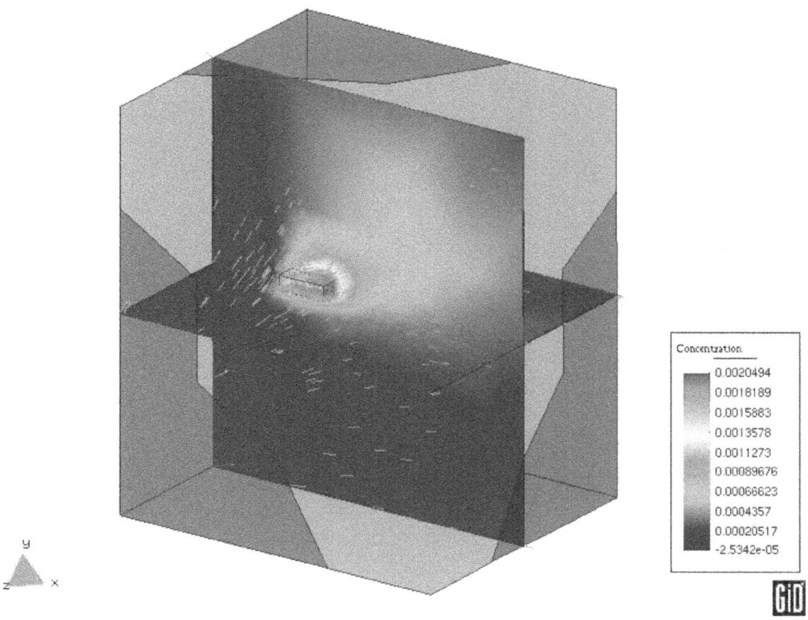

Figure 6.29: Normalized concentration distribution of DCA in limestone around mine repository after 600 years.

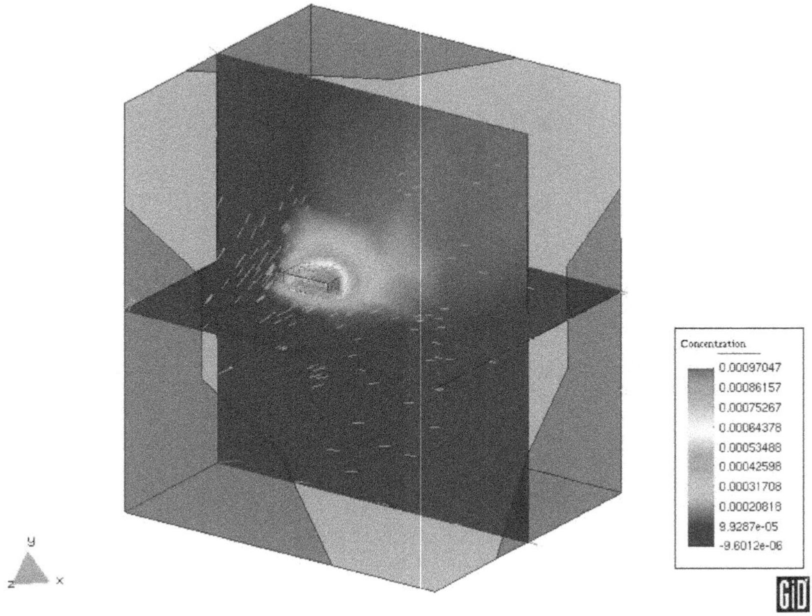

Figure 6.30: Normalized concentration distribution of DCA in limestone around mine repository after 2000 years.

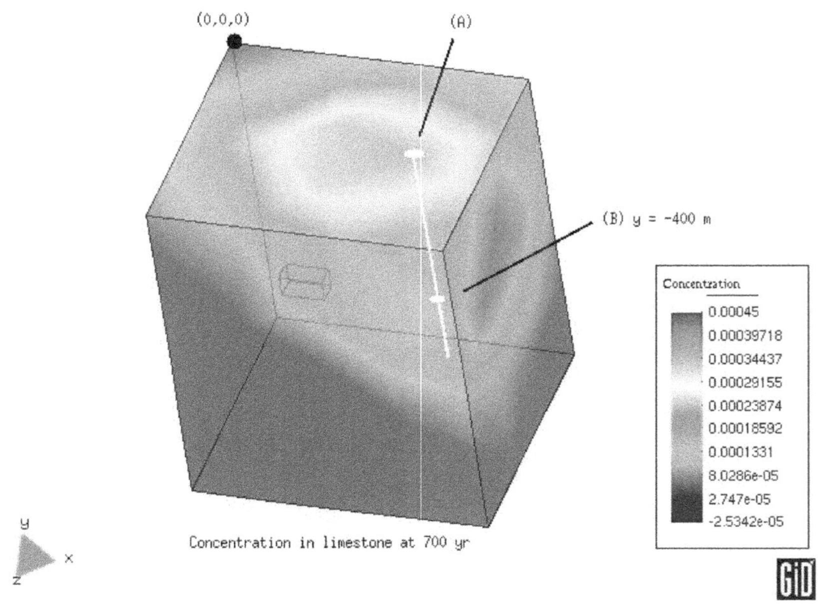

Figure 6.31: Normalized concentration distribution of DCA in limestone around mine repository after 700 years.

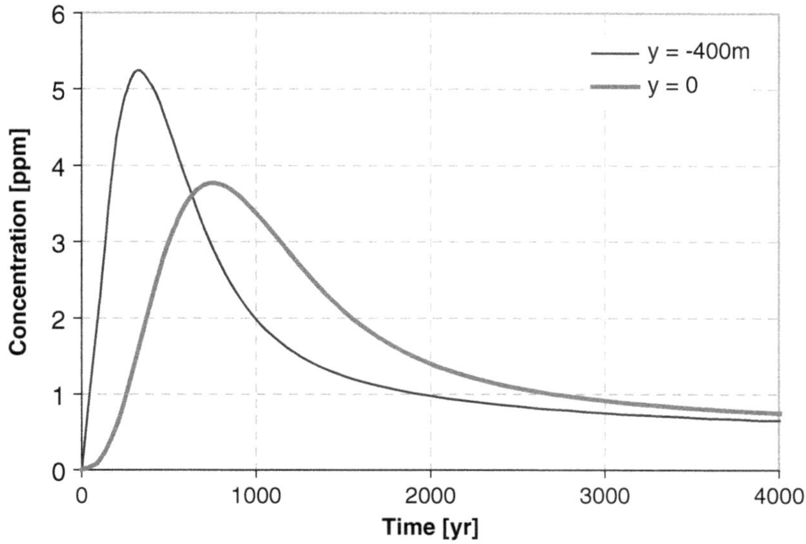

Figure 6.32: Concentration in function of time in the observation well for DCA in limestone (blue: $x = 535$ m; $y = -400$ m; $z = 300$ m) and at the surface (red: $x = 527$ m; $y = 0$ m; $z = 300$ m).

distribution of the fractures and fracture zones. In the case of the limestone the transport through the rock is rapid, compared to crystalline rock, due to much higher hydraulic conductivity.

Figure 6.32 shows the results for short-term analysis of DCA concentrations in two points A and B, see Fig. 6.31. The points A and B are equivalent to the points O_3 and O_2 in Fig. 6.4 for the case of crystalline rock. It can be seen that unlike the case of mine in crystalline rock, here the maximum concentration is reached relatively quickly, after only few hundreds of years, and it decreases from then onward. The maximum concentration is just above 5 ppm reached in approximately 300 years, and drops below 1 ppm in both points after 4000 years.

However, these results are obtained for a very conservative case and in a real case it is very unlikely that the concentrations would reach such level, because of the factors mentioned before, which are for convenience repeated here again:

1. Not all of the dichlorvos would transform/decay into DCA.
2. There would be decay of DCA.
3. Fully saturated flow was considered.
4. It was considered that the clay is saturated at $t = 0$ years. As mentioned previously, practically important wetting of the volume of clay-embedded waste will not commence until 4000 years after application, see Fig. 6.12. This is for the case when the impact of pressure-induced wetting is not taken into

account and providing that the waste mass in the big room in this case is surrounded by a 100 cm 'liner' of Friedland Ton with the assumed density. Therefore, at time $t = 3000$ years the chemical would have not left the repository at all, or very little would have leaked.

6.5.8 Case of disposal of zinc in mine repository in limestone

Second analysis for mine repository in limestone was performed for the case of waste containing batteries. In the following sections the safety aspect of disposal of batteries in abandoned underground mines will be discussed.

6.5.8.1 Modelling conditions for zinc

The geometry of the mine and considered domain and all the parameters in the model remained the same as for DCA except the ones mentioned below.

It was considered that the batteries are mixed with Friedland Ton clay in ratio 50 : 50 by volume. It was assumed that at $t = 0$ years the clay is fully saturated and the batteries are decayed in a way that the zinc can travel through the clay and into the surrounding rock. It was further assumed that 25% of the batteries by weight is zinc. In the case of zinc it was taken that K_d is 0.02 m^3/kg, and the corresponding retardation factor becomes $R = 121$.

6.5.8.2 Transport results for zinc

The same velocity field as in the case of dichlorvos/DCA is obtained for zinc, shown in Figs 6.22–6.26. Figure 6.33 shows the results for zinc concentrations in two points A and B, see Fig. 6.31. It can be seen that the concentrations

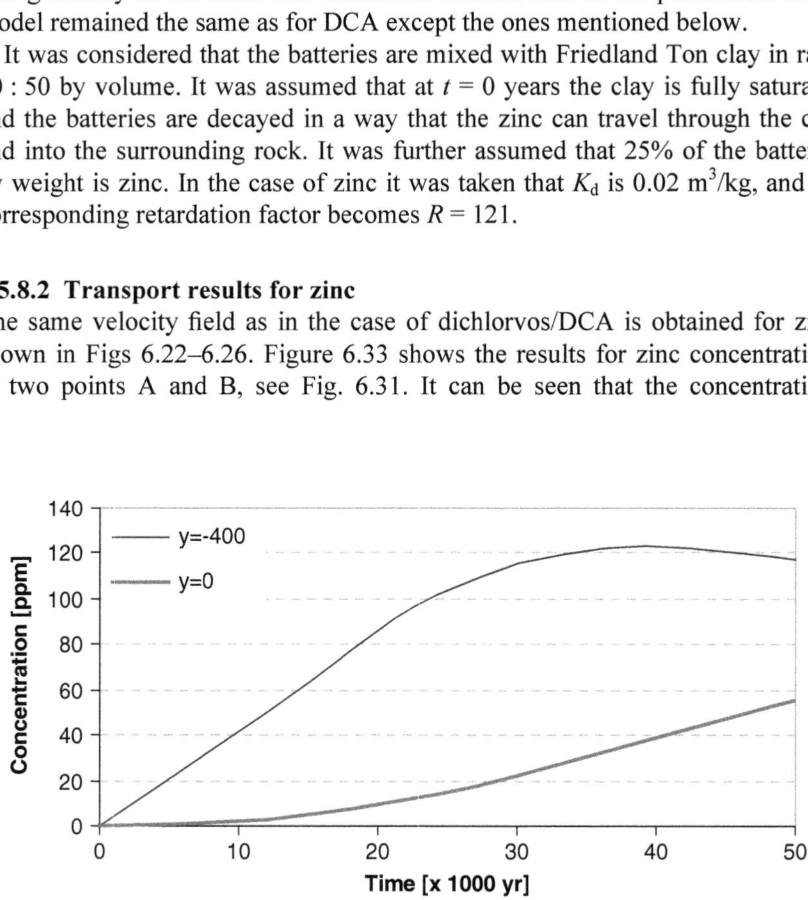

Figure 6.33: Concentration of zinc as a function of time in the observation well (blue: $x = 535$ m; $y = -400$ m; $z = 300$ m) and at the surface (red: $x = 527$ m; $y = 0$ m; $z = 300$ m).

are higher than in the case of disposal of batteries in mine repository in crystalline rock.

However, these results are obtained for a very conservative case and in a real case it is very unlikely that the concentrations would reach such level, because of the following factors:

1. In the analysis it was assumed that the clay is saturated at $t = 0$ years.
2. Fully saturated flow is considered.
3. It is considered that the batteries have decayed at $t = 0$ in such way that zinc is free to travel through the clay and surrounding rock.
4. It is not considered that the rock would absorb and immobilize part of the zinc, which would reduce the amount of zinc available for transport. The influence of the retardation factor R in the equations is to slow down the transport, it does not immobilize the zinc.

6.6 Risk assessment summary

The long-term analysis showed that the hazardous waste embedded in clay of low hydraulic conductivity is slowly released in time. It can be seen that both effects in the transport of the chemicals are evident, the advection and the dispersion in the rock, while the transport in the clay is mainly due to diffusion. The timescale of the whole process of release of the chemical from the repository is the order of 600,000 years. This is what is referred to as long-term analysis in this chapter. The short-term analysis is of the order of few thousand years.

The analysis of mine repository in crystalline rock shows that very low concentrations of chemicals would appear in the groundwater not far from the mine repository and on the surface. In the case of DCA the concentrations on the surface do not exceed 10 ppm before $t \approx 200{,}000$ years.

However, these results are obtained for a very conservative case and in reality it is very unlikely that the concentrations would reach such level, because of the following factors: (i) not all of the dichlorvos would transform/decay into DCA; (ii) there would be decay of DCA and it may be that no DCA ever reach the surface; (iii) fully saturated flow was considered; (iv) the flow velocity vector has got a significant upward (towards the surface) component, which in reality may not be the case.

It must be mentioned that any analysis longer than few thousand years is unreliable since by that time tectonic as well as glacial processes may completely change the situation.

The short-term analysis of DCA leakage and transport shows that the concentration of DCA in the observation well would not exceed 1 ppm in 3000 years. This result is valid for very conservative case, since in reality there would be several factors that would reduce the concentration in the observation well, e.g. it was considered that the clay is saturated at $t = 0$ years. In the case of the Friedland Ton clay with density at water saturation of 1900 kg/m^3, practically important wetting of the volume of clay-embedded waste will not commence until 4000 years after application. This is valid if don't take into account the impact of pressure-induced wetting and providing that the waste mass in the repository is surrounded by a 100 cm 'liner' of Friedland Ton with the assumed density, which was not the case in this study. Therefore, at time $t = 3000$ years the chemical would have not left the repository at all, or very little would have leaked.

The analysis for disposal of batteries in mine repository in crystalline rock show the effects due to the retardation of the zinc during the transport through the surrounding rock, effect which did not exist in the case of DCA. Similar propagation pattern to the one found for dichlorvos/DCA is obtained for zinc, with the difference of slower propagation rate. The results of analysis show that in the period of 120,000 years the concentration of zinc in the observation well will not exceed 2 ppm. However, these results are obtained for a very conservative case and in reality it is very unlikely that the concentrations would reach such level, because of the following factors: (i) in the analysis it was assumed that the clay is saturated at $t = 0$ years; (ii) fully saturated flow is considered; (iii) it is considered that the batteries have decayed at $t = 0$ in such way that zinc is free to travel through the clay and surrounding rock; (iv) it is not considered that the rock would absorb and immobilize part of the zinc, which would reduce the amount of zinc available for transport.

The overall risk assessment of disposal of hazardous chemicals similar to dichlorvos or batteries in mine repositories in crystalline rock show that the risk in short-term, few thousand years, are minimal for the groundwater and surrounding environment, providing that a proper engineered barrier is implemented. The long-term analysis is in favour of the approach as well, however, because of the reasons mentioned above, any results giving predictions for more than 1000 years must be taken with caution.

There are similarities and differences in the processes of transport of DCA once it leaves the repository, depending on whether the repository is in crystalline rock or limestone. Both processes are similar in the process of release of the chemicals from the repository, the main mechanism for transport being diffusion in clay. The differences are in respect to the transport in the surrounding geologic media. In crystalline rock the main transport is conducted through fractures and fracture zones, since there the hydraulic conductivity is much higher than in the rock. The crystalline rock slows down the transport by absorbing the chemical, which penetrates the rock mainly by diffusion. In the case of repository in crystalline rock the transport will be mainly be defined by the characteristics and distribution of the fractures and fracture zones. In the case of the limestone the transport through the rock is rapid, compared to crystalline rock, due to much higher hydraulic conductivity and is due to both, advection and dispersion.

The results for short-term analysis of DCA concentrations show that the maximum concentration in the observation well is reached relatively quickly, after only few hundred years, and it decreases from then onward. The maximum concentration is just above 5 ppm reached in approximately 300 years, and drops below 1 ppm after 4000 years.

These results, just like the ones in the case of mine repository in crystalline rock, are obtained for a very conservative case and in reality it is very unlikely that the concentrations would reach such level, because of the factors mentioned above.

For the case of disposal of batteries in mine repositories in limestone it can be seen that the concentrations are higher than in the case of disposal of batteries in mine repository in crystalline rock, with maximum concentrations reaching over 120 ppm in 40,000 years.

However, these results are obtained for a very conservative case and in reality it is very unlikely that the concentrations would reach such level, because of the factors mentioned above.

The above risk analysis shows that the disposal in mine repositories in limestone should also represent a safe option providing that the engineered barrier can provide sufficiently high insulation. The analysis shows that the engineered barrier in the case of mine repository in limestone is more important than in the case of crystalline rock because of the ways of transport of chemicals through these two different types of geological media, which has been described in this chapter.

References

[1] Bear, J., Tsang, C.-F. & de Marsily, G. (eds.), *Flow and Contaminant Transport in Fractured Rock*, Academic Press, Inc.: San Diego, 1993.

[2] Bear, J. & Berkowitz, B. Groundwater flow and pollution in fractured rock aquifers. *Development of Hydraulic Engineering*, Vol. 4, ed. P. Novak, Elsevier Applied Science: Oxford, 1987.

[3] Adler, M.P. & Thovert, J.-F. (eds.), *Theory and Applications of Transport in Porous Media. Fracture and Fracture Networks*, Vol. 15, Kluwer Academic Publishers: Dordrecht, 1999.

[4] Barenblatt, G.I., Zheltov, I.P. & Kochina, I.N., Basic concepts in the theory of homogeneous liquids in fissured rocks. *Journal of Applied Mathematics and Statistics*, **24**, pp. 1286–1303, 1960 [in Russian].

[5] Gerke, H.H. & van Genuchten, M.T., A dual-porosity model for simulating the preferential movement of water and solutes in structured porous media. *Water Resources Research*, **29(2)**, pp. 305–319, 1993.

[6] Warren, J.E. & Root, P.J., The behaviour of naturally fractured reservoirs. *Society of Petroleum Engineers Journal*, **3**, pp. 245–255, 1963.

[7] Odeh, A.S., Unsteady-state behaviour of naturally fractured reservoirs. *Society of Petroleum Engineers Journal*, pp. 60–64, March 1965.

[8] Snow, D.T., Anisotropic Hydraulic conductivity of Fractured Media. *Water Resources Research*, **5(6)**, pp. 1273–1289, 1969.

[9] Brown, S.R. & Scholz, C.H., Closure of random elastic surfaces in contact. *Journal of Geophysical Research*, **90**, pp. 5531–5545, 1985.

[10] Gentier, S., Morphologie et Comportement Hydromechanique d'une Fracture Naturell dans une Granite sous Contrainte Normale PhD Thesis, Universite d'Orleans, 1986.

[11] Brown, S.R., Fluid flow through rock joints; the effect of surface roughness. *Journal of Geophysical Research*, **92(B2)**, pp. 1337–1347, 1987.

[12] Moreno, L., Tsang, C.F., Tsang, Y. & Neretnieks, I., Some anomalous features of flow and solute transport arising from fracture aperture variability. *Water Resources Research*, **26(10)**, pp. 2377–2391, 1988.

[13] Moreno, L., Tsang, Y.W., Tsang, C.F., Hale, F.V. & Neretnieks, I., Flow and transport in a single fracture: A stochastic model and its relation with field observations. *Water Resources Research*, **24(12)**, pp. 2033–2048, 1988.

[14] Barton, N., Bandis, S. & Bakhtar, K. Strength, deformation and conductivity coupling of rock joints. *International Journal of Rock Mechanics and Mining Sciences*, **22**, pp. 121–140, 1985.

[15] Mourzenko, V.V., Thovert, J.-F. & Adler, P.M., Permeability of a single fracture; validity of the Reynolds equation. *Journal de Physique II*, **5**, pp. 465–482, 1995.

[16] Dienes, J.K., Permeability, percolation and statistical crack mechanism. Issues in Rock Mechanics. *Proc. 22nd Symp. on Rock Mechanics*, Berkley, University of California, August 1987.

[17] Long, J.C.S., Remer, J.S., Wilson, C.R. & Witherspoon, P.A., Porous media equivalents for networks of discontinuous fractures. *Water Resources Research*, **18**, pp. 645–658, 1982.

[18] Cacas, M.C., Ledoux, E., Marsily, G.D., Tillie, B., Barbeau, A., Durand, E., Feuga, B. & Peaudecerf, P., Modeling fracture flow with a stochastic discrete fracture network: Calibration and validation. 1. The flow model. *Water Resources Research*, **26**, pp. 479–489, 1990.

[19] Mercer, J.W. & Faust, C.R., Geothermal reservoir simulation. 3: Application of liquid-and-vapor-dominated hydrothermal modelling techniques to Wairakei, New Zealand. *Water Resources Research*, **15(3)**, pp. 653–671, 1979.

[20] Charlaix, E., Guyon, E. & Roux, S., Permeability of a random array of fractures of widely varying apertures. *Transport in Porous Media*, **2**, pp. 31–43, 1987.

[21] Robinson, P.C., Numerical calculations of critical densities for lines and planes. *Journal of Physics A: Mathematical and General*, **17(14)**, pp. 2823–2830, 1984.

[22] Charlaix, E., Guyon, E. & Rivier, N., A criterion for percolation threshold in a random array of plates. *Solid State Communication*, **50(11)**, pp. 999–1002, 1984.

[23] Wilke, S., Guyon, E. & de Marsily, M., Water penetration through rocks: test of a three-dimensional percolation description. *Mathematical Geology*, **17(1)**, pp. 17–24, 1985.

[24] Alboin, C., Jaffré, J., Joly, P., Roberts, J.E. & Serres, C., A comparison of methods for calculating the matrix block source term in a double porosity model for contaminant transport. *Computational Geosciences*, **6**, pp. 523–543, 2002.

[25] Brebbia, C.A., Telles, J.C. & Wrobel, L.C., *Boundary Elements Techniques*, Springer-Verlag: Berlin, 1984.

[26] Zagar, I. & Skerget, L., Integral formulations of a diffusive-convective transport equation, BE applications in Fluid Dynamics, eds. C.A. Brebbia & H. Power, Computational Mechanics Publications: Southampton, Boston, pp. 153–176, 1995.

[27] Nardini, D. & Brebbia, C.A., A new approach to free vibration analysis using boundary elements. *Applied Mathematical Modelling*, 7, pp. 157–162, 1983.

[28] Partridge, P.W., Brebbia, C.A. & Wrobel, L.C., *The Dual Reciprocity Boundary Elements Method*, Computational Mechanics Publications: Southampton, Boston, 1992.

[29] Golberg, M.A. & Chen, C.S., The theory of radial basis functions applied to the BEM for inhomogeneous partial differential equations. *Boundary Elements Communications*, **5**, pp. 57–61, 1994.

[30] Popov, V. & Power, H., DRM-MD approach for the numerical solution of gas flow in porous media with application to landfill. *Engineering Analysis Boundary Elements*, **23**, pp. 175–188, 1999.

[31] Popov, V., Power, H. & Baldasano, J.M., BEM solution of design of trenches in a multi-layered landfill. *Journal of Environmental Engineering*, **124/1**, pp. 59–66, 1998.

[32] Popov, V. & Power, H., Numerical analysis of the efficiency of landfill venting trenches. *Journal of Environmental Engineering*, **126/1**, pp. 32–38, 2000.

[33] Popov, V. & Power, H., The DRM-MD integral equation method: an efficient approach for the numerical solution of domain dominant problems. *International Journal of Numerical Methods in Engineering*, **44**, pp. 327–353, 1999.

[34] Florez, W.F. & Power, H., DRM multidomain mass conservative interpolation approach for the BEM solution of the two-dimensional Navier-Stokes equations. *Computers & Mathematics with Applications*, **43(3–5)**, pp. 457–472, 2002.

[35] Florez, W.F. & Power, H., Multi-domain mass conservative dual reciprocity method for the solution of the non-Newtonian Stokes equations. *Applied Mathematical Modelling*, **26/3**, pp. 397–419, 2002.

[36] Florez, W.F., *Nonlinear Flow Using Dual Reciprocity*, WIT Press: Southampton, 2001.

[37] Samardzioska, T. & Popov, V., Numerical comparison of the equivalent continuum, non-homogeneous and dual porosity models for flow and transport in fractured porous media. *Advances in Water Resources*, **28**, pp. 235–255, 2005.

[38] Natalini, B. & Popov, V., Tests of radial basis functions in the 3D DRM-MD *Communications in Numerical Methods in Engineering*, **22(1)**, pp. 13–22, 2005.

[39] Peratta, A. & Popov, V., A new scheme for numerical modelling of flow and transport processes in 3D fractured porous media. *Advances in Water Resources*, **29(1)**, pp. 42–61, 2006.

[40] Peratta, A., BEM applied to Flow and Transport in Fractured Porous Media, PhD Thesis, Wessex Institute of Technology, Southampton, UK, University of Wales, December 2004.

[41] Saad, Y. SPARSKIT, *A Basic Tool Kit for Sparse Matrix Computations*. Technical Report RIACS-90-20, Research Institute for Advanced Computer Science, NASA Ames Research Center: Moffett Field, CA, 1990.

[42] US EPA, Phase I Comments for Dichlorvos (D238925; Case No. 819293; Chemical No. 084001), Memorandum-Peer review of DDVP, 1999 http://www.epa.gov/oppsrrd1/op/ddvp/efedrisk.pdf

[43] Macler, B.A. & Pontius, F.W., Update on the groundwater disinfection rule. *Journal of the American Water Works Association*, **89(1)**, pp. 16–20, 1997.

[44] WHO, Disinfection of Water, Local Authorities, Health and Environment, Briefing pamphlet series, No. 3 (1995).

[45] Simpson, K.J. & Hayes, K.P., Drinking water disinfection by-products: An Australian perspective. *Water Research*, **32(5)**, 1522–1528, 1998.

[46] Compilation of physical and physico/chemical data of clay materials and steel containers that are suitable for waste isolation, LowRiskDT Project, D2.2 Report, March 2002.

Appendix to Chapter 2

A2.1 Austria

A2.1.1 Active mines and mineral production

Mines in Austria and their annual production for 1998 are shown in Table A2.1.

Table A2.1: Mines in Austria (based on the US Geological Survey).

Mineral	Operating companies	Name of the mines/ location	Annual production (10^3 tons)
Coal	Graz-Koflacher Eisenbahn und Bergbaugesellschaft GmbH (Government 100%)	Oberdorf Mine	1,200
Graphite	Industrie und Bergbaugesellschaft Pryssok & Co KG	Trandorf Mine at Móhldorf	15
Graphite	Grafitbergbau Kaiserberg Franz Mayr-Melnhof & Co	Kaisersberg Mine	3
Graphite	Grafitbergbau Trieben GmbH	Trieben Mine	3
Gypsum	Erste Salzburger Gipswerk-Gesellschaft Christian Moldan KG	Abtenau and Moosegg Mines	300
Gypsum	Rigips Austria GmbH	Grundlsee, Puchberg, Unterkainisch, and Weisenbach Mines	250
Gypsum	Knauf Gesellschaft GmbH	Hinterstein Mine	160
Iron ore	Voest-Alpine Erzberg GmbH (Government 100%)	Erzberg Mine at Eisenerz	2,000
Magnesite	Veitsch-Radex	AG Mines at Breitenau, Hochfilzen and Radenthein	600
Magnesite	Radex Austria AG (Osterreichische Magnesit AG 100%)	Millstatteralpe Mine	250
Talc	Luzenac Naintsch AG	Mines at Lassing, Rabenwald, and Weisskirchen, Plants at Oberfeistitz and Weisskirchen	160
Tungsten	Wolfram Bergbau und Hótten GmbH Mittersill	Mine, Salzburg; conversion plant, Bergla	350

A2.1.2 Inactive mines

No specific information could be retrieved on inactive mines of the country except Schmitzbe coal mine, which closed in 1995, and Trimmelkam, which closed in 1992.

A2.2 Belgium

A2.2.1 Active mines and mineral production

Mineral production in Belgium for the years 1996–1998 is presented in Table A2.2.

Table A2.2: Production of industrial minerals in Belgium (based on the US Geological Survey).

	Production (10^3 tons, unless otherwise specified)		
Mineral	1996	1997	1998
Dolomite	3,379	3,466	3,500
Limestone	33,000	30,000	30,000
Petit granite (Belgian bluestone) (m^3)	1,200,000	1,200,000	1,000,000
Sodium sulphate	250	250	250

The country has been an important producer of marble for more than 2000 years. All the marble quarries are in Wallonia. Active mines and quarries in Belgium and their annual production for 1998 are shown in Table A2.3.

Table A2.3: Active mines and quarries in Belgium (based on the US Geological Survey).

Mineral	Operating companies	Name of the mines/location	Annual production (10^3 tons)
Dolomite	SA Dolomeuse (Group Lhoist)	Quarry at Marche les Dames	500
Dolomite	SA de Marche-les-Dames (Group Lhoist)	Quarries at Namèche	3,000
Dolomite	SA Dolomies de Merlemont (Group Lhoist)	Quarry at Philippeville	100
Limestone	Carmeuse S.A. (Long View Investment NV)	Mines at Engis	1,850

continued

Table A2.3: *Continued.*

Mineral	Operating companies	Name of the mines/location	Annual production (10^3 tons)
Limestone	Carmeuse S.A. (Long View Investment NV)	Mines at Frasnes	450
Limestone	Carmeuse S.A. (Long View Investment NV)	Mines at Maizeret	850
Limestone	Carmeuse S.A. (Long View Investment NV)	Mines at Moha	800
Limestone	SA Transcar (Royal Volker Stevin)	Mines at Maizeret	850

A2.2.2 Inactive mines

Very little information has been retrieved about inactive mines in Belgium. The only abandoned mines found are some coal mines, located throughout the country. These mines are presented in Table A2.4.

Table A2.4: Inactive mines in Belgium.

Name	Mineral exploited	Location	Dates of Operation
Le Hasard (Cheratte) (underground mine)	Coal	Liége	1860s–1977
Blegny-Trembleur	Coal	Liége	Closed until the mid 1980s
Bas Bois	Coal	Liége	Closed until the mid 1980s
Houthalen	Coal	Kempen	Closed until 1992
Winterslag	Coal	Kempen	Closed until 1992
Andre Dumont	Coal	Waterschei (Kempen)	Closed until 1992
Eisden	Coal	Kempen	Closed until 1992
Kleine Heide	Coal	Beeringen (Kempen)	Closed until 1992
Voort	Coal	Zolder (Kempen)	Closed until 1992
Monceau-Fontaine 14	Coal	Charleroi	Closed until the mid 1980s
Marcinelle Nord	Coal	Charleroi	Closed until the mid 1980s

continued

Table A2.4: *Continued.*

Name	Mineral exploited	Location	Dates of Operation
Bois du Cazier	Coal	Charleroi	Closed until the mid 1980s
St Catherine	Coal	Charleroi	Closed until the mid 1980s
Anderlues	Coal	Centre	Closed until the mid 1980s
St Albert in Ressaix	Coal	Centre	Closed until the mid 1980s
Bois du Luc in Houdeng Aimeries	Coal	Centre	Closed until the mid 1980s

Note: Due to lack of available information, it is not possible to determine which of the above are underground mines, except 'Le Hasard' mine.

A2.3 Denmark

A2.3.1 Active mines and mineral production

Denmark has no known economically exploitable reserves of metallic ores, so the mining activity is concentrated in industrial minerals. Tables A2.5 and A2.6 show the production of industrial minerals and active mines, respectively.

Table A2.5: Production of industrial minerals (based on the US Geological Survey).

Mineral	Production (tons unless otherwise specified)		
	1996	1997	1998
Chalk	359,378	427,634	425,000
Clays (e)			
Fire clay	1,800	20 (*)	20
Kaolin	3,000	3,000	2,500
Other	8,050	8,000	6,000
Moler, extracted (thousand cubic meters)	185	185	185
Lime, hydrated and quicklime	108,628	115,129	116,000
Salt, all forms	600,000 (*)	600,000 (e)	600,000

continued

Table A2.5: *Continued.*

Mineral	Production (tons unless otherwise specified)		
	1996	1997	1996
Sand and gravel (e)			
Onshore (thousand cubic meters)	18,000	18,000	18,000
Offshore (thousand cubic meters)	5,000	5,000	5,000
Stone			
Dimension (mostly granite) (e)	27,198 (*)	26,000	26,000
Limestone			
Agricultural	695,380	700,000 (e)	700,000
Industrial (e)	250,000	250,000	250,000

Note: Table includes data available through March 1999 based on estimated sales of domestically produced mineral commodities; * reported production; e, estimated.

Table A2.6: Active mines in Denmark (based on the US Geological Survey).

Mineral	Major operating companies and major equity owners	Location of main facilities	Annual capacity (10^3 tons)
Chalk	A/S Faxe Kalkbrud	Quarries at Stevns and Sigerslev	250
Diatomite (moler) (thousand cubic meters)	Dansk Moler Industri A/S (Damolin)	Quarries on Mors and Fur Islands	145
Kaolin	Aalborg Portland A/S	Mine and plant on Bornholm Island	25
Salt	Dansk Salt I/S	Mine (brine) at Hvornum, plant at Mariager	600

A2.3.2 Inactive mines

No specific information could be retrieved on inactive mines of the country.

A2.4 Finland

A2.4.1 Active mines and mineral production

The ore output of Finnish mines between 1944 and 1999 is shown in Fig. A2.1.

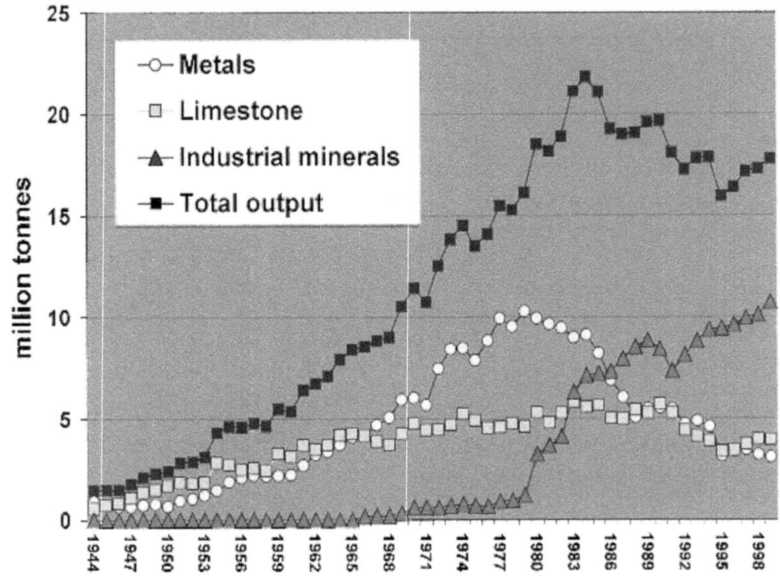

Figure A2.1: Ore output of Finnish mines between 1944 and 1999 (based on the Geological Survey of Finland).

The metallic ore mines in Finland for 1998 are shown in Table A2.7.

Table A2.7: Metallic ore mines (based on the US Geological Survey).

Mineral	Major operating companies and major equity owners	Location of main facilities	Annual capacity (10^3 tons)	Mine type
Chromite	Outokumpu Oyj (Government, 40%; Insurance Co., 12.3%)	Mine at Kemi	1,000	OP+UG
Copper: Ore, Cu content	Outokumpu Oyj (Government, 40%; Insurance Co., 12.3%)	Mines at Pyhasalmi, Saattopora, and Hitura	10	UG
Gold: Ore, Au content (tons)	Outokumpu Oyj (Government, 40%; Insurance Co., 12.3%)	Mine at Orivesi	4	UG
Gold: Ore, Au content (tons)	Williams Resources Inc.	Pahtavaara Mine near Sodankyla	3	OP

continued

Table A2.7: *Continued.*

Mineral	Major operating companies and major equity owners	Location of main facilities	Annual capacity (10^3 tons)	Mine type
Nickel: Ore, Ni content	Outokumpu Oyj (Government, 40%; Insurance Co., 12.3%)	Mine at Hitura	3	UG
Zinc: Ore, Zn content	Outokumpu Oyj (Government, 40%; Insurance Co., 12.3%)	Mine at Pyhasalmi	25	UG

OP, open-pit; UG, underground mining.

In addition, limestone mines and industrial mineral mines according to the Geological Survey of Finland for 1999 are shown in Tables A2.8 and A2.9, respectively.

Table A2.8: Limestone mines (based on the Geological Survey of Finland).

Mine	District	Mineral	Owner	Total ore output (tons)	Mine type
Parainen	Parainen	Lms	Partek Nordkalk	1,279,870	OP+UG
Ihalainen	Lappeenranta	Lms, Wol	Partek Nordkalk	1,172,826	OP
Putkinotko	Vampula	Dol	Partek Nordkalk	151,834	OP+UG
Ruokojärvi	Kerimäki	Lms, Dol	Partek Nordkalk	251,838	UG
Ryytimaa	Vimpeli	Dol	Partek Nordkalk	184,816	OP
Tytyri	Lohja	Lms	Partek Nordkalk	197,367	UG
Förby	Särkisalo	Lms	Karl Forsström	170,225	UG
Siikainen	Siikainen	Dol	Partek Nordkalk	91,517	OP
Sipoo	Sipoo	Dol, Lms	Partek Nordkalk	158,670	UG
Kalkkimaa	Tornio	Dol	Saxo Minerals	92,012	OP
Ankele	Virtasalmi	Dol	Saxo Minerals	72,078	OP
Reetinniemi	Paltamo	Dol	Juuan Dolomiittikalkki	34,820	OP
Vesterbacka	Vimpeli	Lms	Partek Nordkalk	22,865	OP
Mustio	Karjaa	Lms	Partek Nordkalk	20,819	OP
Matara	Juuka	Dol	Juuan Dolomiittikalkki	17,345	OP
Siivikkala	Vampula	Dol	Partek Nordkalkk	14,216	OP
Total				3,934,785	

Lms, limestone; Wol, wollastonite; Dol, dolomite; OP, open-pit; UG, underground mining.

Table A2.9: Industrial mineral mines (based on the Geological Survey of Finland).

Mine	District	Mineral	Owner	Total ore output (tons)	Mine type
Siilinjärvi	Siilinjärvi	Ap, Lms	Kemira Chemicals	8,818,542	OP
Horsmanaho	Polvijärvi	Tlc, Ni	Mondo Minerals	491,651	OP
Lahnaslampi	Sotkamo	Tlc, Ni	Mondo Minerals	589,441	OP
Kinahmi	Nilsiä	Qz	SP Minerals	182,534	OP
Lipasvaara	Polvijärvi	Tlc, Ni	Mondo Minerals	58,013	OP
Kemiö	Kemiö	Qz, Fsp	SP Minerals	52,570	OP
Ristimaa	Tornio	Qz	Saxo Minerals	54,982	OP
Haapaluoma	Peräseinäjoki	Fsp	SP Minerals	0	OP
Total				10,247,733	

Ap, apatite; Lms, limestone; Tlc, talc; Fsp, feldspar; Qz, quartz; OP, open-pit.

Figure A2.2 shows the location of the major active mines in Finland.

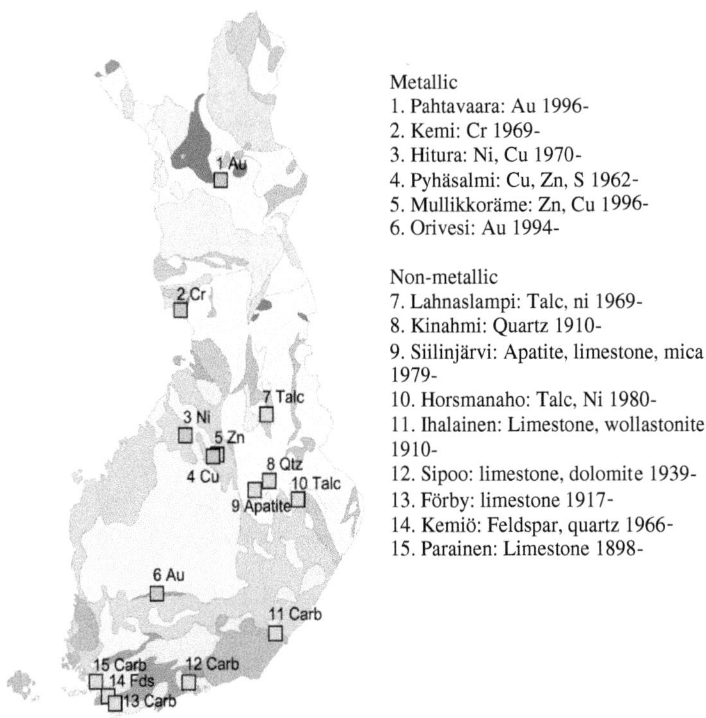

Figure A2.2: Major active mines in Finland (based on the Geological Survey of Finland).

Figure A2.3 shows the active and inactive gold mines and the significant proven and prospective gold deposits in Finland, while the industrial mineral mines and quarries in Finland are shown in Fig. A2.4.

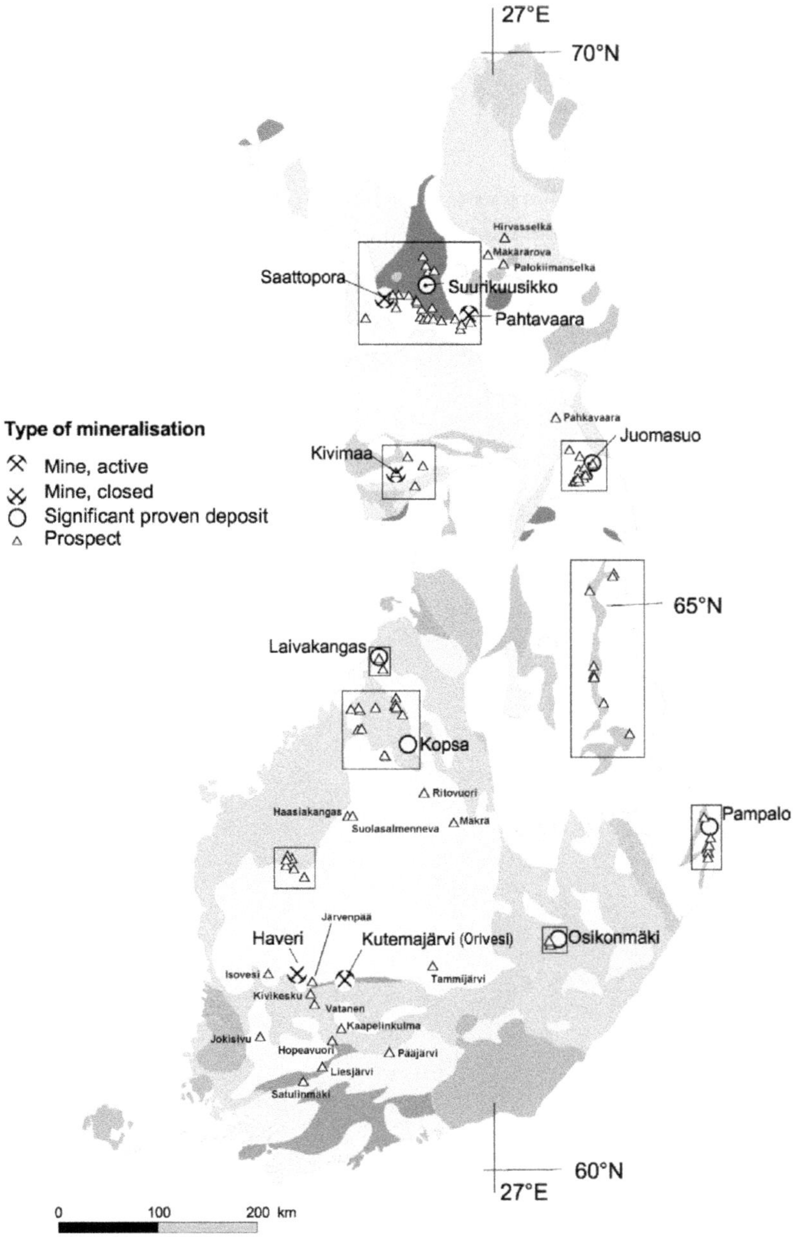

Figure A2.3: Active and abandoned gold mines and proven and prospective gold deposits in Finland (based on the Geological Survey of Finland).

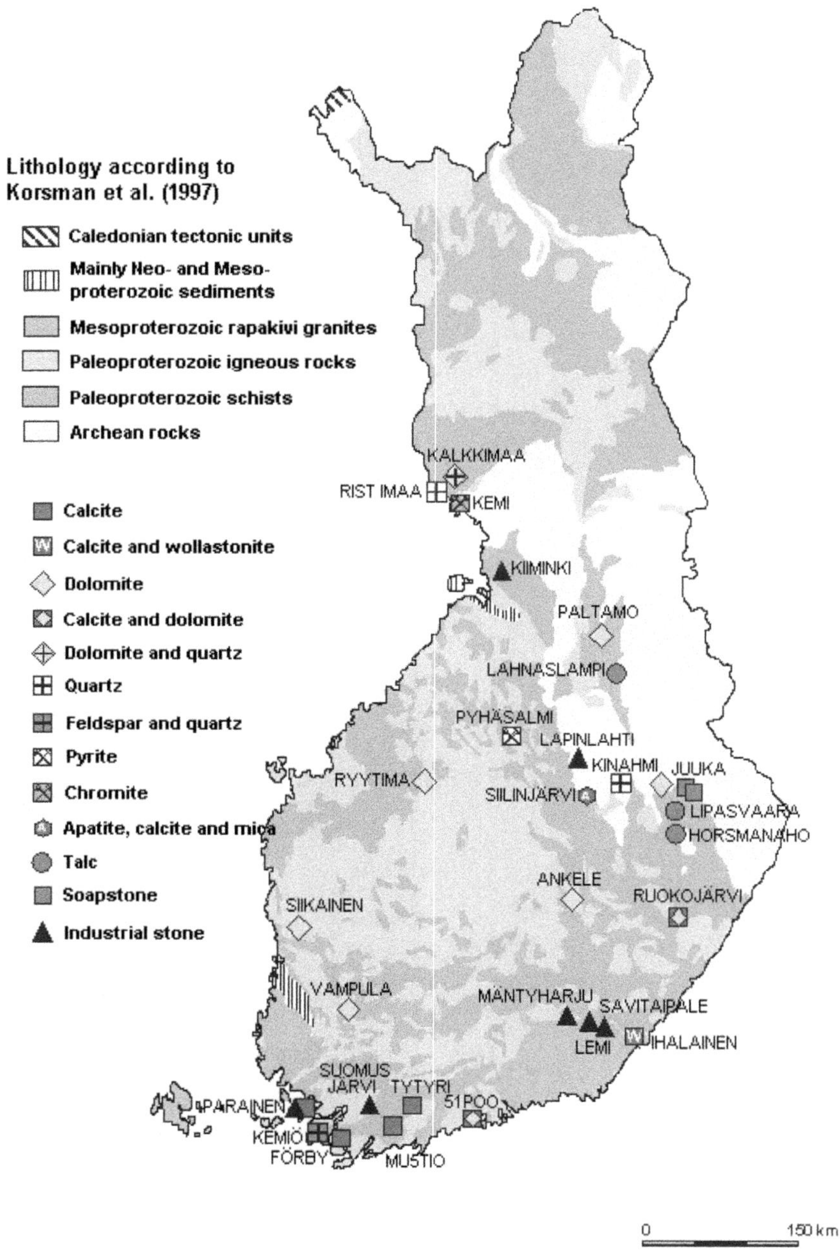

Figure A2.4: Industrial mineral mines and quarries in Finland (based on the Geological Survey of Finland).

A2.4.2 Inactive mines

Information about inactive mines in Finland are included in the websites of Geological Survey of Finland and Outokumpu Oy, which is the leading company in this country. These inactive mines are shown in Table A2.10.

Table A2.10: Inactive underground mines in Finland.

Name	Mineral exploited	Location	Owner	Dates of operation
Enonkoski	Ni–Cu	Enonkoski, Savonlinna	Outokumpu Finnmines Oy	1984–1994
Vammala*	Ni–Cu	Vammala	Outokumpu Finnmines Oy	1974–1994
Kotalahti*	Ni–Cu		Outokumpu Finnmines Oy	1957–1987
Aijala*	Cu–Zn	Orijarvi		1948–1961
Metsamonttu*	Cu–Zn–Pb	Orijarvi		1951–1974
Luikonlahti*	Cu–Zn		Malmikaivos Oy	1958–1983
Vuonos*	Cu–Zn		Outokumpu Finnmines Oy	1967–1968
Vihanti*			Outokumpu Oy	1952–1992
Otanmaki*	Fe		Rautaruukki Oy	1949–1985
Kivimaa	Au	Tervola	none	1969–?
Saattopora	Au	Kittila	Outokumpu Oy	1988–1995
Haveri	Au	Viljakkala	Baltic Minerals Finland Oy	18th century and 1942–1962
Kuurmanpohja*	Al–Fe	Joutseno	Paroc Oy Ab	
Mullikkorame	Cu–Zn	Mullikkorame		1 year remaining
Pyhasalmi	Cu–Zn			5 years remaining

*It cannot be specified whether they are open-pit or underground mines due to lack of information

A2.5 France

A2.5.1 Active mines and mineral production

Active mines in France and their annual production for 1998 are shown in Table A2.11.

Table A2.11: Mines in France (based on the US Geological Survey).

Mineral	Operating companies	Name of the mines/location	Annual production (10^3 tons)
Andalusite	Denain-Anzin Minéraux Refractaire Ceramique (DAMREC)	Glomel Mine, Brittany	75
Barite	Barytine de Chaillac	Mine and plant at Chaillac, Indre Province	150
Barite	Société Industrielle du Centre	Mine at Rossigno, Indre Province	100
Coal	Charbonnages de France (CdF), Bassin de Paris	Mines and washeries in middle France	2,500
Coal	Charbonnages de France (CdF), Bassin de Nord-Pas-de-Calais	Mines and washeries in northern France	1,000
Coal	Charbonnages de France (CdF), Bassin de Lorraine	Mines and washeries in eastern France	9,500
Feldspar	Denain-Anzin Minéraux S.A.	Mine and plant at St. Chély d'Apcher	55
Fluorspar	Société Générale de Recherches et d'Exploitation Minière (SOGEREM)	Mines at Le Burc, Montroc le Moulina, and Trebas	150
Gold	Société des Mines du Bourneix (Government)	Mines in the Saint Yrieix la Perche District, Limoges	4,000 (kg)
Gold	Mines d'Or de Salsigne (Eltin Co., 51%; Co., 18%; Peter Hambro Plc., 10%)	Ranger Mine near Carcassonne	3,000 (kg)
Gypsum	S.A. de Materiel de Construction	Mine at Taverny	1,500
Kaolin	La Source Compagnie Minière	Kaolin d'Arvor Mine, Quessoy	300
Potash, K_2O	Mines de Potasse d'Alsace S.A. (MDPA)	Mines at Amélie, Marie-Louise, and Theodore, in Alsace	10,000
Salt, rock	Compagnie des Salins du Midi et des Salines Varangeville de l'Est	Mine at Saint-Nicolas-de-Port	9,000
Talc	Talcs de Luzenac S.A. (Rio Tinto Corp., 100%)	Trimons Mine near Ariège, Pyrenees	350,000
Uranium, U_3O_8	Compagnie Générale des Matières Nucleaires (COGEMA) (Government)	Mines at Limousin, Vendee, and Hérault	1,800

A2.5.2 Inactive mines

Various inactive mines are presented in Table A2.12.

Table A2.12: Inactive mines in France.

Name	Mineral exploited	Location	Owner	Dates of operation
Ensisheim	Coal		Charbonnages de France	
Ungersheim	Coal		Charbonnages de France	
Rudolphe	Coal		Charbonnages de France	
Marie/ Marie-Louise	Coal		Charbonnages de France	
Staffelfelden	Coal		Charbonnages de France	
Berrwiller	Coal		Charbonnages de France	
Theodore	Coal		Charbonnages de France	
Schoenensteinbach	Coal		Charbonnages de France	
Amelie	Coal		Charbonnages de France	
Max	Coal		Charbonnages de France	
Joseph/Else	Coal		Charbonnages de France	
La Mure	Coal	Isére	Charbonnages de France	Closed in 1997
Carmaux	Coal	Tarn	Charbonnages de France	Closed in 1997
Saint-Bel	Pyrite	Saint-Pierre-la-Palud/Rhône		
Mines of Mulhouse	Potash			
Terres Rouges	Iron	Lorraine	ARBED S.A.	Closed in 1998
	Bauxite	Var Province	Aluminium Péchiney Société Anonyme des Bauxites et Alumines	Closed in 1993

A2.6 Germany

A2.6.1 Active mines and mineral production

Production of major minerals in Germany is shown in Table A2.13.

Table A2.13: Production of major minerals (based on the European Association of Mining Industries).

	Production (10^6 unless otherwise specified)		
Mineral	1996	1997	1998
Coal	47.9	46.5	41.3
Lignite	187.2	177	166.2
Oil	2.6	2.8	2.9
Natural gas (billion m^3)	20.7	20.4	19.9
Potash	34.6	35.9	37.1
Rocksalt	4.9	4.1	n/a
Gravel and sand	402	382	370
Quartz and quartz sand	28	28.1	n/a
Quartzite	1.2	1.5	n/a
Limestone	20.2	21.3	n/a
Gypsum	2.6	2.5	n/a
Feldspar (10^3 tons)	359.7	567.3	n/a
Pegmatite (10^3 tons)	319	635.2	n/a
Kaolin	1.8	1.8	n/a
Bentonite (10^3 tons)	491.3	511	n/a
Graphite (10^3 tons)	2.6	1	n/a
Fluorspar (10^3 tons)	87.6	58	60.9
Barytes (10^3 tons)	218	121	210

In addition, major mines in Germany for 1998 are listed in Table A2.14.

Table A2.14: Major mines (based on the US Geological Survey and Industrial Minerals).

Mineral	Major operating companies and major equity owners	Location of main facilities	Annual capacity (10^3 tons)
Bentonite	Sód-Chemie AG	Gammelsdorf, Bavaria	500
Chalk	Kreidewerke Rugen GmbH	Quarries on Rugen Island	500

continued

Table A2.14: *Continued.*

Mineral	Major operating companies and major equity owners	Location of main facilities	Annual capacity (10^3 tons)
	Four companies, about 27 mines, including		Total
Coal,			72,500
anthracite and	Ruhrkohle AG	14 mines in Ruhr region	(40,000)
bituminous	Saarbergwerke AG	5 mines in Saar basin	(14,000)
	Preussag Anthrazit GmbH	Mine at Ibbenburen	(2,500)
Gypsum	Gebr. Knauf Westdeutsche Gipswerke GmbH	Mines in Bavaria, Hesse, Saarrland, Lower Saxony	2,000
Kaolin	Amberger Kaolinwerke GmbH	Mines at Groppendorf, Hirschau, and Sachsen	100
Limestone	Harz Kalk GmbH	Quarries at Bad Kosen, Rubelaand, and Kaltes Tal	6,000
Lignite	Rheinische Braunkohlenwerke AG (Rheinbraun AG)	Surface mines in Rhenish mining area: Garzweiler, Bergheim, Inden, and Hambach	105,000
Lignite	Lausitzer Braunkohle AG (LAUBAG)	Surface mines in Lausatian mining area: Janschwalde/ Cottbus-Nord, Welzow-Sud, Nochten/Reichswalde	50,000
Potash	Kali und Salz AG	Mines (17) at Bergmannssegen-Hugo, Niedersachen-Riedel, Salzdetfurth, Sigmundshall, Hattorf, Neuhof-Ellers, and Wintershall	4,000
Salt (rock)	Kali und Salz AG	Mines at Bad Friedrichshall-Kochendorf, Braunschweig-Luneburg, Heilbronn, Riedel, Stetten, and Wesel (Borth)	15,000

A2.6.2 Inactive mines

There are a large number of inactive mines located in Germany. Some of them are shown in Tables A2.15–A2.18. It should be specified that there is not much information about their present condition.

Table A2.15: Inactive underground metal mines.

Name	Location/Description	Mineral exploited	Dates of operation
Rammelsberg Mine	Goslar	Lead, zinc, copper	Closed 1989
Erzbergwerk Grund	Bad Grund (Harz/Lower Saxony), Achenbach, Knesebeck, Wiemannsbucht and Meding Shaft	Lead, zinc	Closed 1992
Einheit Mine	Elbingerode (Harz/Lower Saxony), 2 shafts	Pyrite	Closed 1990
Bernard-Koenen 2	Near Sangerhausen (Mansfeld/Sangerhausen)	Copper	Closed 1990
Meggen Mine	Lennestadt (North Rhine-Westfalia), Sicilia- und Baro Shafts	Pyrite, lead, zinc, baryte	Closed 1992
Lüderich mine	Near Bergisch Gladbach (North Rhine-Westfalia), Haupt and Franziska Shaft	Lead, zinc	Closed 1978
Schafberg Shaft	Mechernich/Eifel	Lead	
Türk Shaft	Schneeberg (Erzgebirge/Saxony)	Silver	
Ehrenfriedersdorf Mine	Erzgebirge/Saxony, 2 shafts	Tin	Closed 1991
Altenberg Mine	Erzgebirge/Saxony	Tin	Closed 1991
Damme 2	Damme, (Lower Saxony)	Iron	
Malapertus	Wetzlar, (Sieg / Lahn-Dill)	Iron	
Lower Saxony, Sieg / Lahn-Dill, Waldalgesheim (near Bingen), Oberpfalz (Bavaria) (a large number of remaining headgears of former iron mines)		Iron	

Table A2.16: Inactive underground salt and potash mines.

Name	Location/Description	Mineral exploited	Dates of operation
Mariaglück	Höfer near Celle (Lower Saxony), shafts: *Mariaglück, Habighorst*	Salt and potash	

continued

Table A2.16: *Continued.*

Name	Location/Description	Mineral exploited	Dates of operation
Niedersachsen-Riedel	Hänigsen near Celle/Lerthe (Lower Saxony), shafts: *Niedersachsen* (potash) and *Riedel I* (salt)	Salt and potash	Closed 1997
Bergmannssegen-Hugo	Lerthe (Lower Saxony), shafts: *Hugo, Bergmannsegen*	Potash	Closed 1994
Siegfried	Giesen near Hildesheim (Lower Saxony), shaft: *Siegfried*	Potash	
Hildesia-Mathildenhall	Dieckholzen near Hildesheim (Lower Saxony), shafts: *Hildesia, Mathildenhall*	Potash	
Salzdetfurth	Near Hildesheim (Lower Saxony), *Shafts I, II, III*	Potash	Closed 1992
Glückauf	Sondershausen Südharz (Thüringen), *Shafts I, II, IV, V*	Potash	
Bleicherode	Südharz, Südharz (Thüringen), shafts: *Von Velsen I/II, Kleinbodungen*	Potash	
Sollstedt	Südharz, Südharz (Thüringen), shafts: *Sollstedt, Bernterode I/II*	Potash	
Bischofferode	Südharz, Südharz (Thüringen), shafts: *Bischofferode I/II, Neu-Bleicherode*	Potash	Closed 1993
Springen	Werra (Thüringen/Hessen), shafts *Springen I, II/III, IV/V*	Potash	
Alexandershall	Werra (Thüringen/Hessen), remaining: *Shaft II*	Potash	

Table A2.17: Inactive underground coal mines.

Name	Location	Mineral exploited	Owner	Dates of operation
Eward/Hugo	Ruhr	Coal	Deutsche Steinkohle AG	Closed 2000
Westfalen	Ruhr	Coal	Deutsche Steinkohle AG	Closed 2000
Gottelborn/Reden	Saar	Coal	Deutsche Steinkohle AG	Closed 2000
Wehofen	Duisburg	Coal		

continued

Table A2.17: *Continued.*

Name	Location	Mineral exploited	Owner	Dates of operation
Martin Hoop Colliery	Saxony	Coal		Closed 1977
Anna Colliery	Alsdorf (Aachen)	Coal		Closed in the 1990s
Sophia Jacoba Colliery	Huchelhoven (Aachen)	Coal		Closed in the 1990s
Bochum, Essen, Dortmund, etc. (a large number of closed collieries)		Coal		

Table A2.18: Other inactive underground mines.

Name	Location	Mineral exploited	Dates of operation
Glasebach Shaft	Straßberg, Harz	Fluorspar	Closed 1991
Schönbrunn	Schönbrunn/Vogtland	Fluorspar	Closed 1991
Cäcilia, Hermine und Erna	Closed shafts in Stulln/ Bavaria,	Fluorspar	
Grüberg II	Thülen near Brilon, shaft, closed	Calcspar	
Kropfmühl	Bavaria, 2 shafts	Graphite	Closed 1997
Wilhelm and Schenkenbusch shafts	Witterschlick near Bonn	Clay	Closed since the 1990s
Richard, Gute Hoffnung, Lindenborn, Anton and Niedersachsen shafts	Wirges area near Montabaur	Clay	
Melsbach shaft	near Koblenz (shaft, closed)		
Glückauf and Steiger Shafts	Seilitz-Löthain near Meissen/Elbe		

A2.7 Greece

A2.7.1 Active mines and mineral production

Mineral production in Greece for years 1996–1998 is shown in Table A2.19.

Table A2.19: Mineral production in Greece (based on the European Association of Mining Industries).

Minerals	Annual production (10^3 tons)		
	1996	1997	1998
Alumina	602	584	622
Bauxite	2,452	1,877	1,823
Bentonite, activated and processed	663	735	800
Lignite	59,738	58,939	60,400
Magnesite	682	623	n/a
Magnesia, calcined	119	117	104
Magnesia, dead burned	57	86	100
Nickeliferous ore	2,195	1,887	1,800
Perlite	599	696	n/a
PbS concentrate	11.5	26.1	30
ZnS concentrate	13.6	32.6	39

Active mines in Greece and their annual production for 1998 are shown in Table A2.20.

Table A2.20: Active mines in Greece (based on the US Geological Survey).

Mineral	Operating companies	Name of the mines/location	Annual production (10^3 tons)
Bauxite	Bauxites Parnasse Mining Co. S.A. (Eliopoulos-Kyriakopoulos Group)	Mines at Fokis	2,000
Bauxite	Eleusis Bauxites Mines, S.A. (ELBAUMIN) (National Bank of Greece)	Mines near Drama, Itea, and Fthiotis-Fokis	300
Bauxite	Delphi-Distomon S.A.; Hellenic Bauxites of Distomon S.A.; (Aluminium de Grèce S.A.)	Opencast mines at Delphi-Distomon area	500
Bentonite	Mykobar Mining Co. S.A. (Silver and Baryte Ores Mining Co. S.A.)	Mines at Adamas, Milos Island	180
Bentonite	Silver and Baryte Ores Mining Co. S.A.	Mines at Adamas, Milos Island	500
Bentonite	Mediterranean Bentonite Co. S.A. (Industria Chemica Mineraria S.p.A., Italy)	Surface mines on Milos Island	20

continued

Table A2.20: *Continued.*

Mineral	Operating companies	Name of the mines/location	Annual production (10^3 tons)
Chromite	Financial-Mining-Industrial and Shipping Corp. (FIMISCO) (IRO)	Tsingeli Mines and plant near Volos	25
Gold, Au in concentrate	TVX Hellas (TVX Gold Inc., Canada	Kassandra Mines, Olympiada	25
Gypsum	Lava Mining and Quarrying Co. S.A.	Altsi deposit, Crete Island	250
Gypsum	Titan Cement Co. S.A.		280
Lead, mine, Pb in concentrate	TVX Hellas (TVX Gold Inc., Canada)	Kassandra mines (Olympias and Stratoni), northeast Chalkidiki	
Lignite	Public Power Corporation (Government)	Megalopolis Mine, central Peloponnesus	7,000
Lignite	Public Power Corporation (Government)	Ptolemais Mine, near Kozani	28,000
Magnesite, concentrate	Viomagn-Fimisco Ltd. (Violignit S.A., 65%, Alpha Ventures, 35%)	Mines at Gerorema and Kakavos, at Mantoudhi, northern Euboea Island	250
Magnesite	Grecian Magnesite S.A.	Mine at Yerakini, Chalkidiki	200
Nickel, ore	General Mining & Metallurgical Co. S.A. (LARCO) (IRO)	Aghios Ioannis Mines near Larymna	500
Nickel, ore		Mines at Euboea	2,500
Perlite	Silver and Baryte Ores Mining Co. S.A.	Mines on Kos and Milos Islands	300
Perlite	Otavi Minen Hellas S.A. (Otavi Minen AG, Germany)	Milos Island	150
Perlite	Do. Bouras Co.	Kos Island	50
Pozzolan (Santorin earth)	Lava Mining & Quarrying Co. Ltd. (Heracles General Cement Co. S.A.)	Quarries in Milos	350
Pozzolan	Titan Cement Co. S.A.		300
Zeolite	Silver and Baryte Mining Co. S.A.	Mine at Pendalofos	100
Zinc, mined, Zn in concentrate	TVX Hellas (TVX Gold Inc., Canada)	Kassandra mines (Olympias and Stratoni), northeast Chalkidiki	25

The Greek marble industry plays a leading role in the international dimension stone market, as a result of the marble production in almost all areas of the country, its variety of uses and many colours (ash, black, brown, green, pink, red, and multicoloured) (Fig. A2.5).

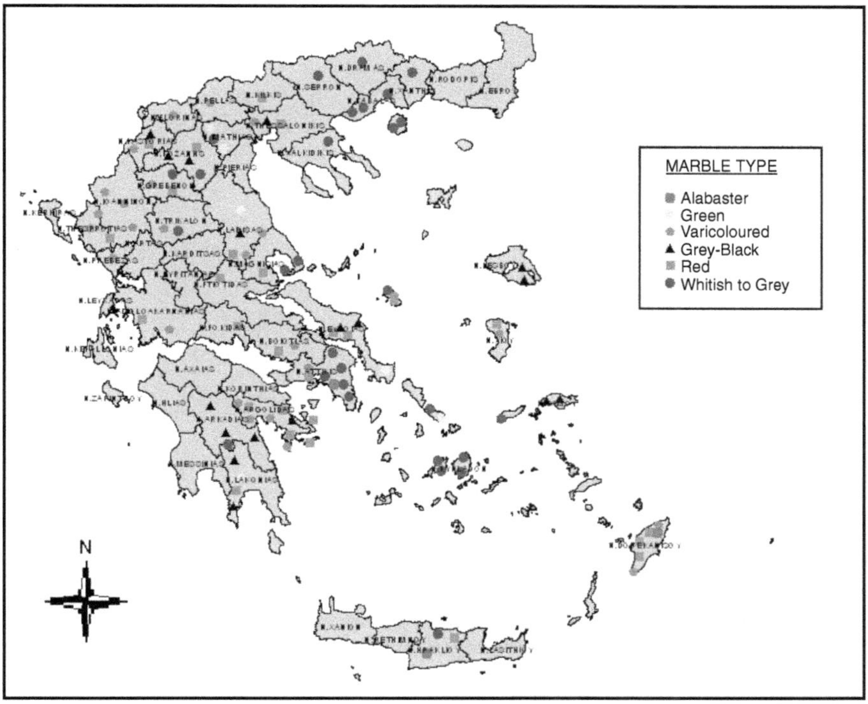

Figure A2.5: Location of marble deposits in Greece.

PPC is the major producer of lignite, the predominant fuel in electricity generation in Greece. PPC continued exploration in the basins of Amyntaion, Elasson, Florina, Megalopolis, and Ptolemais. PPC had reserves estimated to be 6.8 billion tons from which 4 billion tons was estimated to be economically recoverable by open pit mining. Most PPC lignite is produced from the Ptolemais-Amyntaion basin with lesser amounts from the Megalopolis basin (Fig. A2.6).

A2.7.2 Inactive mines

Various inactive mines in Greece are presented in Table A2.21.

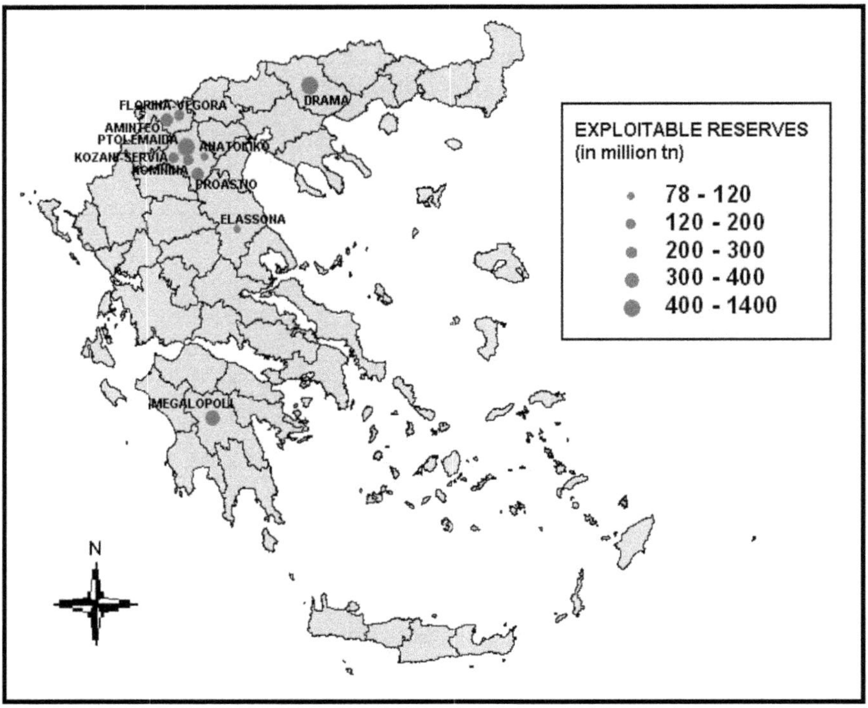

Figure A2.6: Lignite deposits in Greece.

Table A2.21: Inactive mines in Greece.

Name	Mineral exploited	Location	Owner	Exploitation type
Tsagli	Chromite	Eretria		Underground and Open-pit
	Chromite	Domokos	National Bank of Greece	Open-pit
Koromilies 1	Bauxite	Amfissa		Underground and Open-pit
Paliampela	Bauxite	Amfissa	Bauxites Parnasse Mining Co. S.A.	Underground and Open-pit
Psorachi	Bauxite	Amfissa	Bauxites Parnasse Mining Co. S.A.	Underground and Open-pit
Kokkinochoma	Bauxite	Amfissa		Open-pit
Makrilakoma 1	Bauxite	Amfissa		Open-pit
Sideritis	Bauxite	Amfissa	Bauxites Parnasse Mining Co. S.A.	Open-pit

continued

Table A2.21: *Continued.*

Name	Mineral exploited	Location	Owner	Exploitation type
Stifari	Bauxite	Amfissa	Bauxites Parnasse Mining Co. S.A.	Open-pit
Koromilies 2	Bauxite	Amfissa		Open-pit
Koromilies 3	Bauxite	Amfissa		Open-pit
Makrilakoma 2	Bauxite	Amfissa		Open-pit
Zidani	Asbestos	Kozani	Hellenic Mineral Mining Co. S.A.	Open-pit

A2.8 Ireland

A2.8.1 Active mines and mineral production

Table A2.22 shows the mineral production in Ireland for the years 1996–1998.

Table A2.22: Mineral production in Ireland (based on the European Association of Mining Industries).

	Production (10^3 tons unless otherwise specified)		
Mineral	1996	1997	1998
Lead (metal in concentrate)	45.3	45	35.9
Zinc (metal in concentrate)	164.5	193	177.2
Silver ('000 kg in lead concentrate)	14.7	13.3	10.8
Gypsum	422.8	477	500
Alumina	1,233.5	1,272.8	1300
Natural gas (billion m^3)	2.74	2.42	1.79

Today, there are only three active mines in Ireland: the Tara Mine, the Galmoy and Lisheen Mine (Table A2.23).

Table A2.23: Irish-based metal mines (based on Dhonau N.B.).

Name	Minerals	Dates of operation	Type of mine
Navan	Zn, Pb	1977 to at least 2010	Underground
Galmoy	Zn, Pb	1997 to at least 2012	Underground
Lisheen	Zn, Pb	1999 to at least 2015	Underground

A2.8.2 Inactive mines

Some inactive mines in Ireland are shown in Table A2.24 while Fig. A2.7 shows the location of both active and inactive mines.

Table A2.24: Inactive mines in Ireland (based on Dhonau N.B.).

Name	Minerals	Dates of operation	Type of mine
Tynagh	Cu, Zn, Pb, Ag, Ba	1965–1981	Open-pit & Underground
Silvermines	Zn, Pb, Ba	1968–1982	Underground
Gortdrum	Cu, Hg, Ag	1967–1975	Open-pit
Avoca	Cu, Pyrite	1969–1982 (history of mining since 1725)	Open-pit and Underground

A2.9 Italy

A2.9.1 Active mines and mineral production

Major mineral production in Italy for the years 1996–1998 is shown in Table A2.25.

Table A2.25: Major mineral production (based on the European Association of Mining Industries).

	Production (tons unless otherwise specified)		
Minerals	1996	1997	1998
Lead (67% Pb)	21,000	17,600	10,100
Zinc (55% Zn)	20,100	15,400	4,470
Gold	0	0	1.2
Lignite	223,000	203,061	83,700
Oil (10^3 tons)	5,430	5,400	5,600
Natural gas (million N m^3)	20,200	19,500	19,160
Geothermal steam (10^3 tons)	31,000	32,100	34,200
Barytes	80,500	26,300	36,000
Bentonite	475,000	512,900	592,000
Dolomite	781,000	760,000	711,370
Feldspar and aplite (10^3 tons)	2,300	2,200	2,748
Fluorspar	103,000	105,800	107,000
Rocksalt (10^3 tons)	2,941	3,507	3,354
Talc	136,000	141,000	138,000

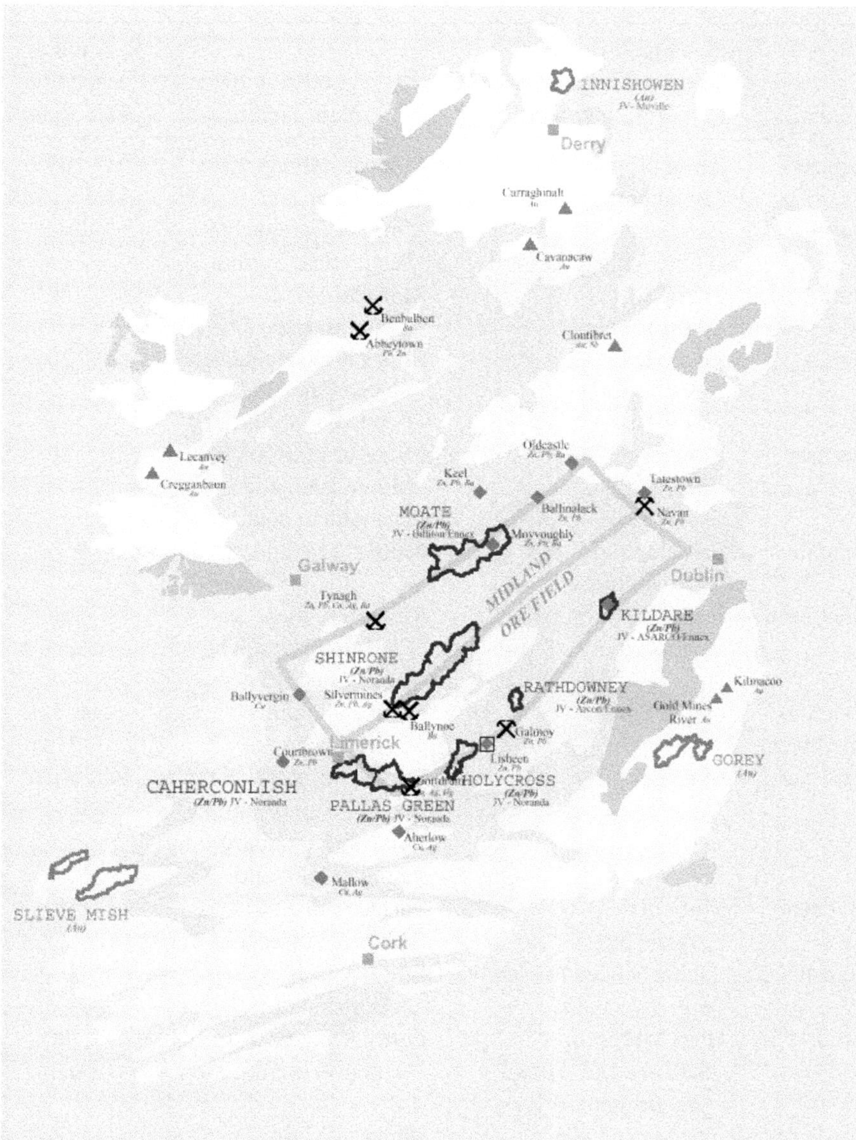

Figure A2.7: Location of present and past mines in Ireland (based on Minco plc.).

Active mines in Italy are shown in Table A2.26.

Table A2.26: Active mines (based on the US Geological Survey and Industrial Minerals).

Mineral	Major operating companies and major equity owners	Location of main facilities	Annual capacity (10^3 tons)
Asbestos	Amiantifera di Balangero S.p.A.	Mine at Balangero, near Turin	100
Barite	Bariosarda S.p.A (Ente Mineraria Sarda)	Mines at Barega and Mont 'Ega, Sardinia	100
Barite	Edem S.p.A. (Government)	Mines at Val di Castello, Lucca	20
Barite	Edemsarda S.p.A. (Soc. Imprese Industriali)	Mines at Su Benatzu, Sto Stefano, and Peppixeddu, Sardinia	20
Barite	Mineraria Baritina S.p.A	Mines at Marigolek, Monte Elto, and Primaluna, near Milan	20
Bauxite	Sardabauxiti S.p.A. (Government)	Mine at Olmedo, Sardinia	350
Bentonite	Industria Chimica Carlo Laviosa S.p.A	Mines and plant on Sardinia Island, and a plant near Pisa	250
Calcium carbonate	Omya S.p.A.	Mine and plant at Carrara, Nocera	Over 500 (1994)
Feldspar	Maffei S.p.A.	Surface mines at Pinzolo, Sondalo, and Campiglia Marittima; underground mine at Vipiteno	(200) (300)
Feldspar	Miniera di Fragne S.p.A.	Surface mine at Alagna Valsesia	(60)
Feldspar	Sabbie Silicee Fossanova S.P.A. (Sasifo)	Surface mine at Fossanova	(30)
Gold	Gold Mines of Sardinia Ltd. 70%, Government 30%	Furtei Mine near Cagliaria, Sardinia	1,400 (kg)
Lead–zinc, ore	Enirisorse S.p.A. (Government)	Mines at Masua, Monteponi, and Sardinia	60
Lignite	Ente Nazional per l'Energia Electtrica (ENEL)	Surface mines at Pietrafitta and Santa Barbara	1,500

continued

Table A2.26: *Continued.*

Mineral	Major operating companies and major equity owners	Location of main facilities	Annual capacity (10^3 tons)
Marble	A number of companies, largest of which include: Mineraria Marittima Srl	Quarries in the Carrara and Massa areas	2,000
Olivine	Nuova Cives Srl.	Mine and processing at Vidracco, Piemonte	300
Potash ore	Industria Sali Otassici e Affini per Aziono S.p.A.	Underground mines at Corvillo, Pasquasia, Racalmuto, and San Cataldo, in Sicily	1,300
Potash ore	Sta Italiana Sali Alcalini S.p.A. (Italkali)	Underground mines at Casteltermini and Pasquasia, Sicily	700
Pumice	Pumex S.p.A.	Quarry, Lipari Island, north of Sicily	600
Pumice	Europumice Srl	Pian di Valle, La Collina, Le Mandarie and S Giovanni delle Contee	150
Pyrite	Nuova Solmine S.p.A.	Underground mines at Campiano and Niccioleta	900
Salt, rock	Sta Italiana Sali Alcalini S.p.A. (Italkahi)	Underground mines at Petralia, Racalmuto, and Realmonte, Sicily	4,000
Salt, rock	Solvay S.p.A.	Underground mines at Buriano, Pontteginori, and Querceto, Tuscany	2,000
Talc	Luzenac Val Chisone S.p.A.	Mines at Pinerolo, near Turin, and at Orani, Sardinia	120
Talc	Talco Sardegna S.p.A.	Mine at Orani, Sardinia	20

A2.9.2 Inactive mines

Some inactive mines in Italy are presented in Table A2.27.

Table A2.27: Inactive mines in Italy.

Name	Mineral exploited	Location	Owner	Dates of operation
Niccioleta	Pyrite	Niccioleta		Closed 1992
	Coal	Sardinia	Carbosulcis S.p.A.	
Pasquasia	Potash	Sicily		Standby
Racalmuto	Potash	Sicily		Standby
Realmonte	Potash	Sicily		Standby

A2.10 Luxembourg

A2.10.1 Mines and mineral production

Mining activity in Luxembourg is very limited and consists of domestic-scale industrial minerals operations. Thus, no specific information could be retrieved on active and inactive mines of the country.

A2.11 Portugal

A2.11.1 Active mines and mineral production

Major mineral production in Portugal for the years 1997 and 1998 are shown in Table A2.28.

Table A2.28: Mineral production in Portugal (based on the European Association of Mining Industries).

	Production (10^3 tons)	
Mineral	1997	1998
Uranium (U_3O_8)	20	22
Iron/manganese	18,905	19,570
Beryllium	3	na
Copper conc. (25% Cu)	444,063	469,172
Tin	6,511	5,594
Tungsten	1,791	1,436
Ornamental rock	1,249,446	na
Industrial rock	86,053,493	na
Pegmatites with lithium	6,838	7,800

continued

Table A2.28: *Continued.*

Mineral	Production (10^3 tons)	
	1997	1998
Salt	595,997	580,209
Feldspathic sands	8,550	9,000
Quartz	9,177	9,000
Feldspar	81,597	80,000
Diatomite	1,540	1,525
Pegmatites (mixed quartz and feldspar)	6,200	6,000
Talc	8,236	8,400

In addition, Fig. A2.8 shows the location of active metallic mines for 1998. Information about the major ones is shown in Table A2.29.

Table A2.29: Major mines in Portugal (based on the US Geological Survey).

Mineral	Major operating companies and major equity owners	Location of facilities	Annual capacity (10^3 tons)
Copper	Sociedade Mineira de Neves-Corvo S.A. (Somincor) (Government, 51%; Rio Tinto Ltd., 49%)	Neves-Corvo Mine near Castro Verde	500
Diatomite	Sociedade Anglo-Portugesa de Diatomite Lda.	Mines at Obidos and Rolica	5
Feldspar	A.J. da Fonseca Lda.	Seixigal Quarry, Chaves	10
Tin	Somincor (Government, 51%; Rio Tinto Ltd., 49%)	Neves-Corvo Mine near Castro Verde	5
Tungsten	Beralt Tin and Wolfram (Portugal) Ltd. (Avocet Mining Plc. 100%)	Panasqueira Mine and plant at Barroca Grande	1,600
Uranium tons	Empresa Nacional de Uranio S.A. (Government 100%)	Mines at Guargia, plant at Urgeirica	150

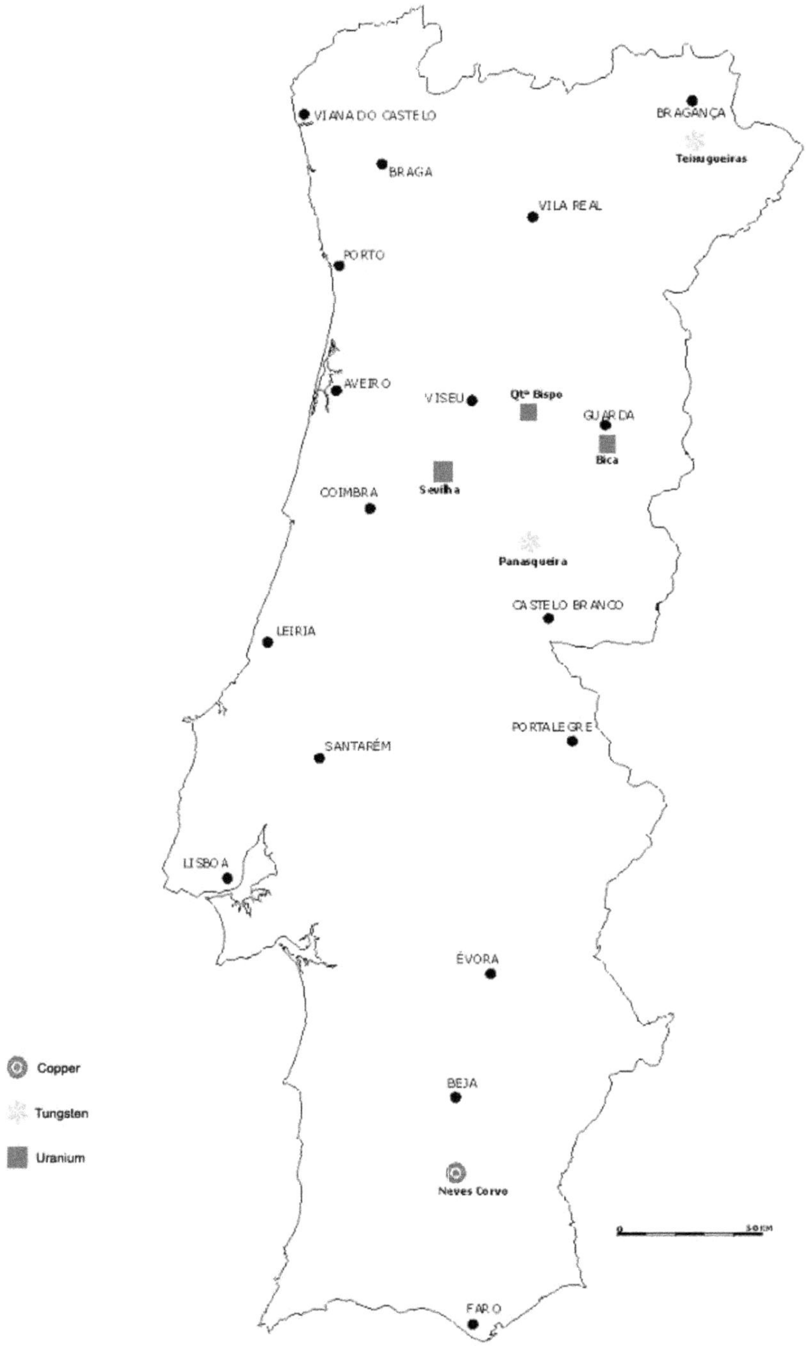

Figure A2.8: Active metallic mines in Portugal (based on the Geological and Mining Institute of Portugal).

Portugal's active industrial mineral mines in 1998 are shown in Fig. A2.9.

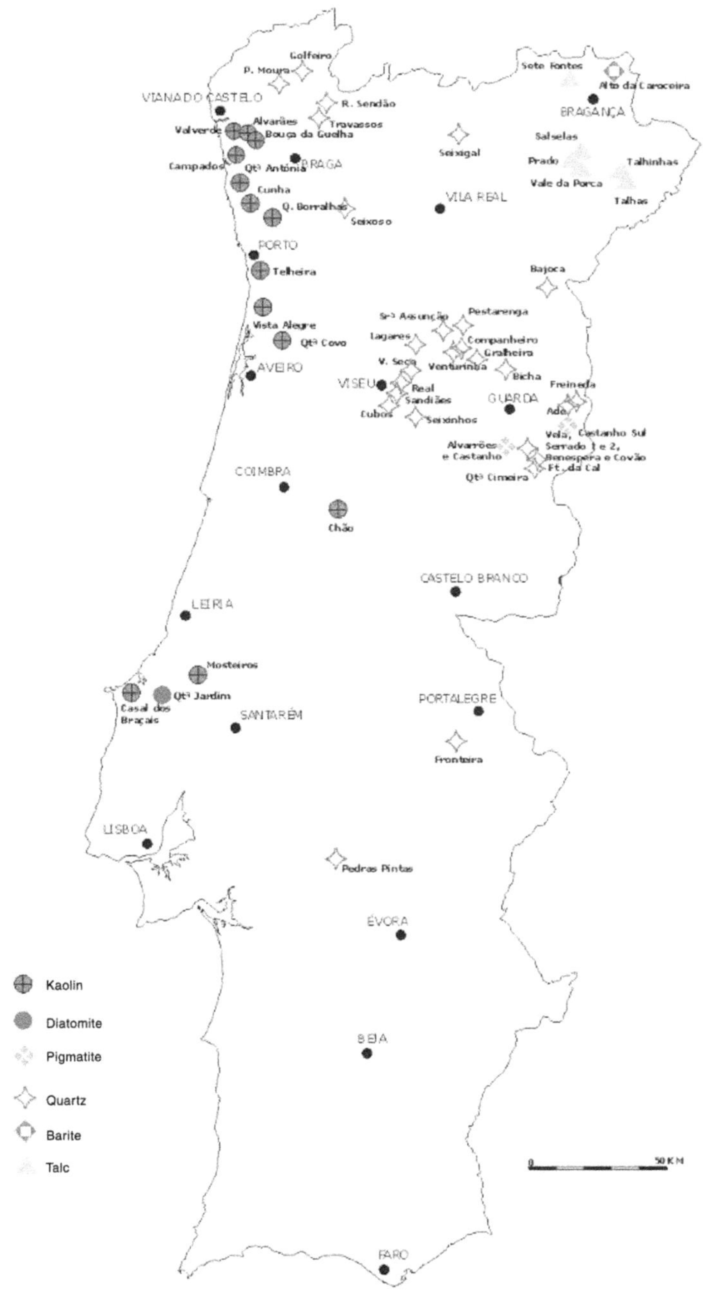

Figure A2.9: Active industrial mineral mines (based on the Geological and Mining Institute of Portugal).

A2.11.2 Inactive mines

The location of some of the country's inactive mines is presented in Fig. A2.10.

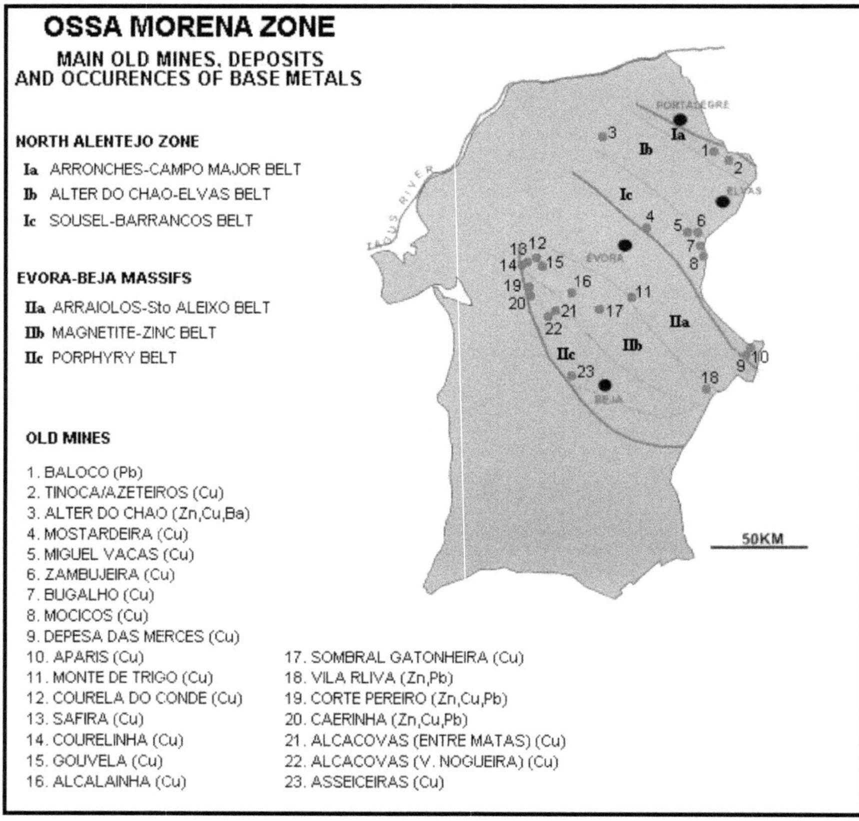

Figure A2.10: Inactive mines (based on the Geological and Mining Institute of Portugal).

A2.12 Spain

A2.12.1 Active mines and mineral production

The mineral production of Spain, from 1996 to 1998, is shown in Table A2.30.

Table A2.30: Mineral production in Spain (based on the European Association of Mining Industries).

Mineral	Production (10^3 tons unless otherwise specified)		
	1996	1997	1998
Non-metallic minerals			
Fluorspar (CaF_2)	117	120	124
Potash (K_2O)	680	639	585
Salt	3,435	3,548	3,620
Quartz	1,438	1,460	1,480
Special clays	1,042	1,460	1,480
Magnesite (MgO)	200	171	170
Sodium sulphate (Na_2SO_4)	859	925	1,001
Celestite ($SrSO_4$)	115	95	111
Washed kaolin	318	296	310
Feldspar	415	398	430
Calcium carbonate	1,650	1,750	1,880
Metallic minerals			
Iron	1,263	58	52
Pyrite	1,042	993	868
Copper (metal content)	38.4	38.4	37.2
Zinc (metal content)	145	147	128
Lead	24	23	19
Gold (kg) (metal content)	2,763	1,824	3,295
Silver (tons) (metal content)	103	66	25
Mercury (tons) (metal content)	861	389	672
Tin (tons) (metal content)	2	4	5
Energy minerals			
Anthracite	6,440	6,678	6,393
Coal	7,195	7,200	6,004
Black lignite	4,071	4,115	3,925
Brown lignite	9,585	8,462	9,750
Oil	513	380	535
Natural gas (million m^2)	466	178	112
Uranium (tons U_3O_8)	346	350	351

In addition, active mines in Spain and their production for 1997 are shown in Table A2.31.

Table A2.31: Active mines in Spain (based on the US Geological Survey and Industrial Minerals).

Mineral	Operating companies	Name of the mines/ location	Annual production (10^3 tons)
Anthracite	Antracitas Gaiztarro S.A.	Mines at Marva and Paulina	2,000
	Antracitas de Gillon S.A.	Mines near Oviedo	2,000
	Antracitas del Bierzo S.A.	Mines near Leon	1,000
	Hulleras del Norte S.A. (Hunosa)	Various mines and plant	3,300
Bituminous	Hulleras Vasco Leonesa S.A.	Santa Lucia Mine, Leon	2,000
	Minas de Figaredo S.A.	Mines near Oviedo	1,000
	Nacional de Carbon del Sur (Encasur)	Rampa 3 and San Jose Mines, Cordoba	200
Lignite	Empresa Nacional de Electricidad (Endesa)	As Pontes Mine, and Andorra Mine, La Coruna	15,000
Barite	Minas de Baritina S.A. (Kali-Chemie of Germany, 100%)	Mine and plant in Espiel area, Cordoba	50
Copper (Ore, metal content)	Atlantic Copper Holding, S.A. (Freeport MacMoRan Inc., 65%, Ercros Group, 35%)	Mines and plant at Arientero, near Santiago de Compostela, Corta Atalay open pit mine.	12
		Cerro Colorado open pit mine and Alredo underground mine, in Rio Tinto area	30
Copper	Navan Resources Ltd.	Migolas and Sotiel areas	6
Fluorspar	Fluoruros S.A. (Bethelhem Steel Corp., 49%)	Opencast mines at San Lino and Val Negro and underground mine at Eduardo, near Carav – all in Asturias	350
	Fluoruros S.A. (Bethelhem Steel Corp., 49%)	Mines at Veneros Sur and Corona, Gijσn	200
Gold	Rio Narcea Gold Mines, Ltd.	Belmonte de Miranda, Asturias	3,750 kg

continued

Table A2.31: *Continued.*

Mineral	Operating companies	Name of the mines/ location	Annual production (10^3 tons)
Iron ore	Compania Andaluza de Minas S.A. (Mokta, 62%)	Mine at Alquife, Granada	4,000
	Altos Hornos de Vizcaya S.A. (U.S. Steel, 25%)	Nine mines in Province of Vizcaya	4,000
	Compania Minera Siderugica de Ponferrada S.A.	Eight mines in Province of Leon	3,000
	Minera del Andevalo S.A.	Opencast mine at Coba, Huelba	2,000
Lead ore	Sociedad Minera y Metalurgica de Penarroya Espana S.A. (Penarroya, France 90%)	Opencast mine at Montos de Los Azules, near Union Murcia	25
	Exploracion Minera International Espana S.A. (EXMINESA)	Underground mine at Rubiales, Lugo	16
	Boliden Apirsa SL	Opencast mine Los Frailes, near Seville	48
Magnesite	Magnesitas de Rubian S.A.	Mines and plant near Sarria, south of Lugo	220
	Magnesitas Navarras S.A.	Mine in Eugui, Navarra	400
Mercury	Minas de Almaden y Arrayanes S.A., (Government, 100%)	Mine and smelter at Almaden	70,000 flasks
Potash, ore	Potasas de Navarra S.A.	Mines and plant near Pamplona	300
	Iberpotasas S.A.	Underground mine at Suria	656
	Union Explosivos Rio Tinto S.A.	Mines at Balsareny/ Sallent and Cardona	2,000
Pyrite	Compania Espanola de Mines de Tharsis	Mines at Tharsis and Zarza, near Seville	1,300
	Rio Tinto Minera S.A. Unvon Explosivos (Rio Tinto, 75%; Rio Tinto Zinc, 25%)	Mines and plant at Rio Tinto, near Seville	900

continued

Table A2.31: *Continued.*

Mineral	Operating companies	Name of the mines/ location	Annual production (10^3 tons)
Sepiolite	Tolsa S.A.	Mine and plant at Vicalvaro, near Toledo	100
	Silicatos-Anglo-Ingleses S.A.	Mine and plant at Villecas near Madrid	200
Uranium, U_3O_8	Empresa Nacional del Uranio (Enusa), (Government,100%)	Mines and plant near Ciudad Real	Metric tons
Zinc Ore	Asturiana de Zinc S.A. (Azsa)	Reocin mines and plants near Torrelavega, Santander	500
	Boliden Apirsa SL	Opencast mine Los Frailes, near Seville	125
	Exploracion Minera International Espana S.A. (EXMINESA)	Underground mine at Rubiales, Lugo	500
	Sociedad Minera y Metalurgica de Penarroya-Espana S.A.	Mines and plants at Montos de los Azules y Sierra de Lujar, San Agustin	200

A2.12.2 Inactive mines

No specific information could be retrieved on inactive mines of the country, except those presented in Table A2.32.

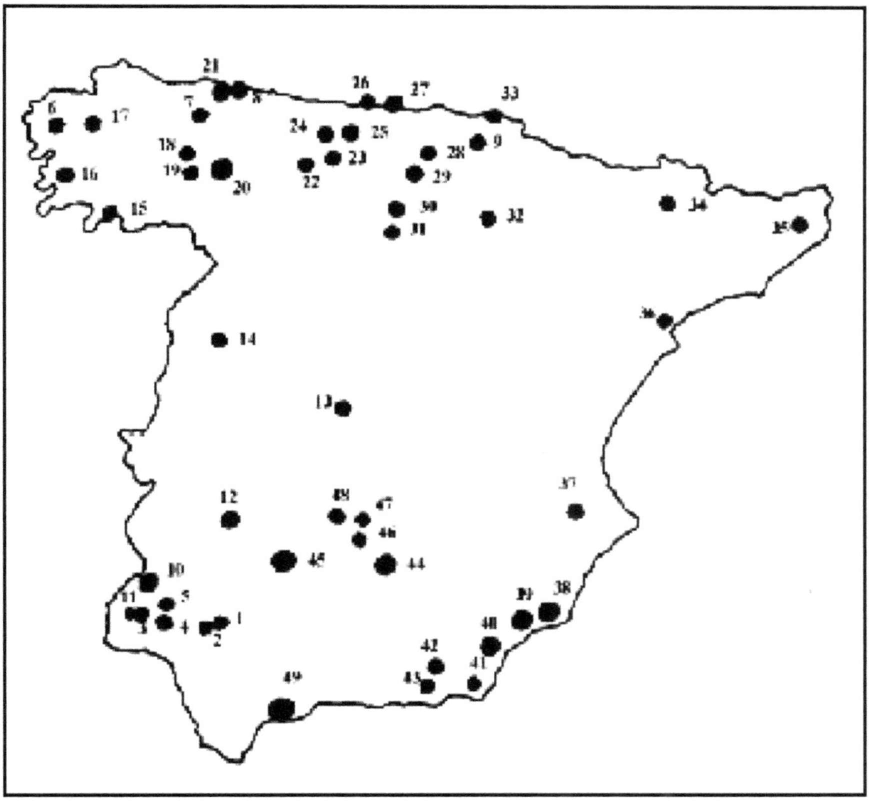

Figure A2.11: Locations of major mining sites and most dangerous tailing ponds in Spain. Key to mines: (1) Los Frailes, (2) Aznalcollar, (3) Tharsis, (4) Sotiel Coronada, (5) Rio Tinto, (6) A Corupa, (7) Belmonte de Miranda, (8) San Juan de Nieva, (9) Mutiloa, (10) Almonaster La Real, (11) Filon sur, (12) Castuera, (13) Rielves, (14) Morille, (15) Xinzo de Limia, (16) Catoira, (17) So-brado, (18) Toreno, (19) Soto y Amio, (20) Carrocera, (21) Avilız, (22) Guardo, (23) Muda, (24) Camalepo, (25) Udias, (26) Suances, (27) Camargo, (28) Maestu, (29) Miranda de Ebro, (30) Valle de Oca, (31) Ibeas de Juarros, (32) Alfaro, (33) Qiarzun, (34) Vilaller, (35) Osor, (36) Bellmunt, (37) Onteniente, (38) Cartegena, (39) Mazar-ron, (40) Cuevad de Almanzora, (41) Nvjar, (42) Almocita, (43) Berja, (44) La Caro-lina, (45) Alcarecejos, (46) Mestanza, (47) Villamayor de Calatrava, (48) Abenojar, (49) Marbella.

Table A2.32: Inactive mines in Spain.

Name	Mineral exploited	Location	Owner	Dates of operation
	Pb-Zn	Underground operation at Reocin	Asturiana de Zinc S.A. (Azsa)	Production is expected to cease in 2003
Troya	Pb-Zn	The Basque Country, Northern Spain	Exminesa (Exploración Minera Internacional España S.A.)	
	Fluorspar	Villabona, near Gijon		Closed in 1992
Aznalcollar	Pb-Zn	Open pit mine at Seville	Boliden Apirsa SL	Operation terminated in 1996
Lieres*	Coal	Nalon, Asturias		Closed in 1999
Mosquitera*	Coal	Nalon, Asturias		
San Vicente*	Coal	Nalon, Asturias		
Entrego*	Coal	Nalon, Asturias		
San Mames, Cerezal*	Coal	Nalon, Asturias		
Olloniego*	Coal	Caudal, Asturias		
Barredo*	Coal	Caudal, Asturias		
Polio*	Coal	Caudal, Asturias		
San Victor*	Coal	Caudal, Asturias		
Santa Barbara*	Coal	Caudal, Asturias		
Entrago*	Coal	Near Oviedo, Asturias		Closed in the 1990s
Herrera	Coal	Sabero, Asturias		

*It cannot be specified whether they are open-pit or underground mines due to lack of information.

A2.13 Sweden

A2.13.1 Active mines and mineral production

Figure A2.12 shows the location of active mines in Sweden and Table A2.33 shows mineral production in Sweden for years 1996–1998.

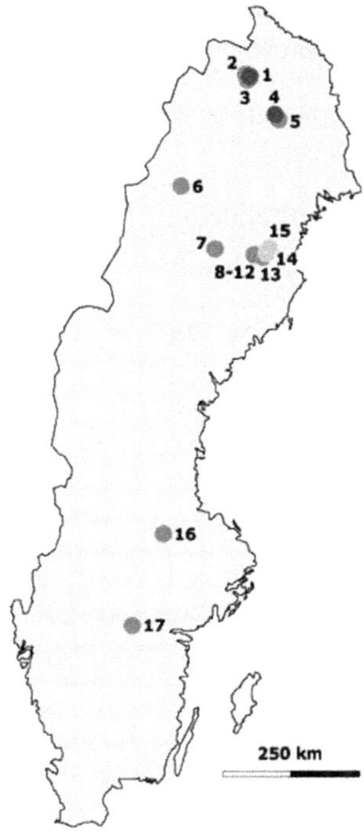

Name of deposit	Operator	Metal	Production (10^3 tons/year)	Type
1. Kiruna	LKAB	Fe	20,000	U
2. Pahtohavare (RC)	Viscaria AB	Cu, Au	290	U
3. Viscaria (RC)	Viscaria AB	Cu	600	U
4. Malmberget	Viscaria AB	Fe	12,000	U
5. Aitik	Boliden AB	Cu, (Au)	18,000	O
6. Laisvall	Boliden AB	Pb, (Zn)	1,950	U
7. Kristineberg	Boliden AB	Cu, Zn, Pb, Au, Ag	560	U
8. Kedtrask (IM)	Boliden AB	Zn	130	O
9. Petiknas	Boliden AB	Cu, Zn, Pb, Au, Ag	440	U
10. Renstrom	Boliden AB		174	U
11. Kankberg (IM)	Boliden AB		113	U
12. ?kulla Ostra (RC)	Boliden AB	Au, Cu, Ag	130	O
13. Langdal (RC)	Boliden AB	Cu, Zn, Pb, Au, Ag	229	O
14. Bjorkdal (RC)	Williams Resources Inc.	Au	1,000	O
15.Akerberg (IM)	Boliden AB	Au	160	U
16. Garpenberg	Boliden AB	Cu, Zn, Pb	930	U
17.Zinkgruvan	North Ltd.	Zn, Pb, Ag	690	U

RC, recently closed; IM, intermittently mined; O, open-pit; U, underground.

Figure A2.12: Active mines in Sweden (based on the Geological Survey of Sweden).

Production of Swedish industrial minerals in 1997 is shown in Table A2.34 while major industrial mineral mines are shown in Table A2.35.

Table A2.33: Production of minerals in Sweden (based on European Association of Mining Industries).

	Production (10^3 tons unless otherwise specified)		
Mineral	1996	1997	1998
Iron ore products	21,228	21,893	20,930
Processed sulphide ores	24,902	23,895	24,182
Copper concentrate	269	315	270
Lead concentrate	136	146	155
Zinc concentrate	292	284	297
Gold in concentrate (tons)	6.1	6.7	5.9

Table A2.34: Production of Swedish industrial minerals in 1997 (based on Industrial Minerals).

Mineral	Production (10^3 tons)
Dolomite, limestone, lime	8,000
Silica sand, quartz	375
Quartzite	260
Clays	200
Diabase	115
Olivine	100
Feldspar, talc, graphite, etc.	80

Table A2.35: Major industrial mineral mines (based on the US Geological Survey and Industrial Minerals).

Mineral	Major operating companies and major equity owners	Location of main facilities	Annual capacity (10^3 tons)
Feldspar	Berglings Malm & Mineral AB (Omya GmbH)	Mines at Beckegruvan, Hojderna, and Limbergsbo	50
Feldspar	Forshammar Mineral AB (Cape Minerals AS)	Mines at Limberget and Riddarhyttan	30
Feldspar	Larsbo Kalk AB (Pluess-Staufer AB)	Mines at Glanshamar and Larsbo	20
Graphite	Woxna Graphite AB (Tricorona Mineral AB, 100%)	Mine and plant at Kringeltjärn, Woxna	20

continued

Table A2.35: *Continued.*

Mineral	Major operating companies and major equity owners	Location of main facilities	Annual capacity (10^3 tons)
Kyanite	Svenska Kyanite AB (Svenska Mineral, 100%)	Quarry at Halskoberg	10
Limestone	Kalproduktion Storugns AB (Nordkalk AB, 100%)	Mines at Gotland Island	3,000
Marble (m^3)	Borghamnsten AB	Quarry at Askersund	15,000

A2.13.2 Inactive mines

According to the Swedish authorities on underground exploitation, the total number of abandoned mines is 25 in Northern Sweden and 775 in central and Southern Sweden. The location of some inactive underground mines is shown in Fig. A2.13. Additional information about many of them can be found in Table A2.36.

Table A2.36: Inactive underground mines in Sweden.

Name	Mineral exploited	Location	Owner	Dates of operation
Stripa	Fe	Stripa	Stripa Mine Service AB	15th century–1977?
Viscaria	Cu		Viscaria AB	Recently closed
Langdal	Au, Zn, Cu		Boliden AB	Recently closed
Pahtohavare	Cu, Au		Viscaria AB	Recently closed
Grangesberg	Fe			Closed 1989
Dannemora	Fe			13th century–closed 1992
Lainejaur	Ni	12 km North of Mala		1941–1945
Enasen		Gavleborg		
Luossavaara	Fe		LKAB	
Tuollavaara			LKAB	
Svappavaara	Fe		LKAB	
Adak	Cu	Adak	Swedish Government	1933–1998?
Laver	Cu	Norrbotten	Boliden AB	1936–1946
Rakkejaur	Zn, Au, Ag			Closed 1988

continued

Table A2.36: *Continued.*

Name	Mineral exploited	Location	Owner	Dates of operation
Åkerberg	Au			1989–(temporarily closed 1999)
Rävliden	Zn, Cu, Pb, Au			Closed 1991
Udden	Zn, Cu, Pb, Au			Closed 1990
Boliden	Au, Cu, Zn, Pb, Ag			Closed 1967
Långsele	Zn, Cu, Pb, Au			Closed 1991
Långdal	Zn, Pb, Au, Ag			1967–recently closed
Kedträsk	Zn			Intermittently mined 1998
Åkulla Östra	Cu, Au, Ag			1997–recently closed
Falun	Cu, Zn, Pb, Au			Closed 1992
Stekenjokk	Zn, Cu, Pb			Closed 1988
Sala	Pb, Zn, Ag			Closed 1962

Other inactive mines with no additional information available are:

Northern district: Brännmyra, Rutjebäcken, Näsliden, Holmtjärn, Kimheden, Hornträskviken, Rävliden, Rävlidmyran, Kankberg, Åkulla västra, Åsen, Östra Högkulla.

Southern district: Smålands Taberg (iron), Hohults Mangangruva (manganese), Jakobsbergs Mangangruva (manganese), Kleva Nickelgruva (nickel), Ädelfors Guldgruva (gold), Sunnerskogs Koppargruva (copper), Rolfsby Stora Mangangruva (manganese), Gustavs Mangangruva (manganese), Storgruvan, Vretgruvan (manganese), Hedvigs Zink och Blygruva (zinc, lead), De Beschiska Koppargruvan (copper), Börgeltorps zinkgruva (zinc), Skälö Koppargruva (copper), Bjuvs Stenkolsgruva (coal, clay), Onslunda Gruvor (calcium fluoride), Långbans Gruvor (sulpide minerals), Getö Stora Silvergruva (silver), Hällefors Östra Silvergruva (silver), Åmmebergs Zinkgruvor (zinc), Stråssa gruvfält (iron, sulphide minerals?), Ljusnarsbergsfältet (iron, sulphide minerals?), Stripås Koppargruva (copper), Storgruvan (sulphide minerals?), Falu Koppargruva (copper, silver), Sågmyra Koppargruva (copper), Storgruvan i Furboberget (iron?), Garpenbergs Odalfält (iron).

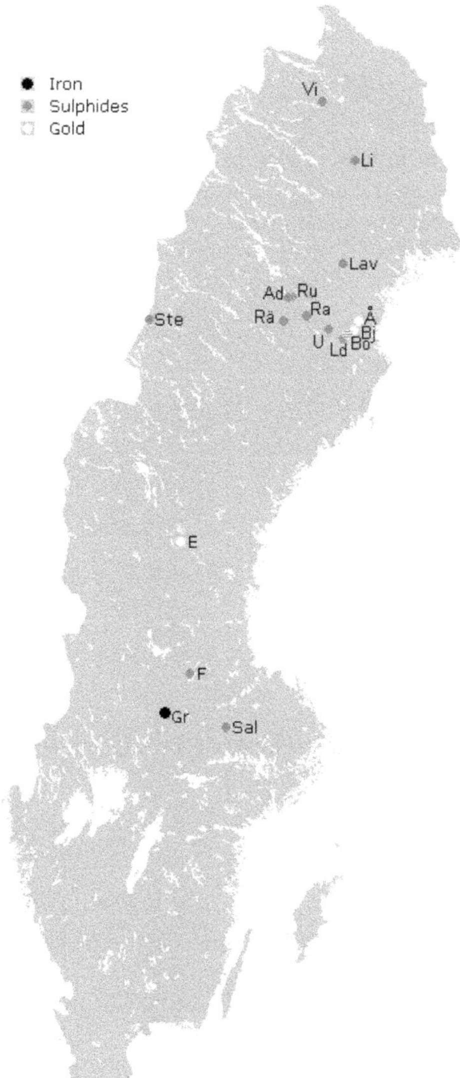

Norrbotten:
Vi= Viscaria, Li= Liikavaara,
Lav= Laver

Skellefte Field:
Ad= Adak, Ra= Rakkejaur,
Å=Åkerberg,
Rä= Rävliden, U= Udden, Bo= Boliden,
Ls= Långsele, Ld= Långdal

Bergslagen:
F = Falun, Gr= Grängesberg, Sal= Sala

Other areas:
Ste= Stekenjokk, E = Enåsen

Figure A2.13: Location of inactive underground mines in Sweden (based on CM Tracing).

A2.14 The Netherlands

A2.14.1 Active mines and mineral production

The Netherlands has no commercially exploitable reserves of metal ores. The only active mines that exist in the country extract industrial minerals. Active mines in the Netherlands and their annual production for 1998 are shown in Table A2.37.

Table A2.37: Mines in the Netherlands (based on the US Geological Survey).

Mineral	Operating companies	Name of the mines/location	Annual production (10^3 tons)
Limestone	Ankerpoort NV (Lhoist SA, 100%)	Mines at Maastricht and Winterswijk	600
Salt	Akzo Salt and Basic Chemicals BV	Mines at Hengelo	2,000
Salt	Akzo Salt and Basic Chemicals BV	Mines at Delfzijl	2,000

A2.14.2 Inactive mines

No specific information could be retrieved on inactive mines of the country.

A2.15 The United Kingdom

A2.15.1 Active mines and mineral production

Major mineral production in UK for the years 1996–1998 is shown in Table A2.38.

Table A2.38: Major mineral production in the UK (based on the European Association of Mining Industries).

	Production (10^3 tons unless otherwise specified)		
Mineral	1996	1997	1998
Coal	50,196	48,495	41,276
Natural gas (oil equivalent)	84,618	86,350	90,467
Crude petroleum (including condensates)	130,007	128,205	132,602
Tin/lead/zinc/iron	5.1	5.2	3.2
China/ball clay (sales)	3,161	3,216	3,364
Other clays and shale	12,483	11,795	12,394

continued

Table A2.38: *Continued.*

	Production (10^3 tons unless otherwise specified)		
Mineral	1996	1997	1998
Limestone and dolomite	103,119	105,034	106,000
Chalk (GB only)	9,239	9,550	9,500
Sandstone	17,522	18,499	18,700
Silica sand	4,861	4,704	4,600
Sand/gravel (land/marine)	96,377	98,383	100,000
Igneous rock	50,903	48,771	49,000
Gypsum	2,000	2,000	2,000
Rock salt	2,200	1,800	700
Brine salt	4,812	4,861	4,800
Fluorspar	65	64	63
Barytes	93	74	68
Potash (KCl)	1,030	941	1,014

In addition, active mines, their owners and annual production for 1998 are shown in Table A2.39.

Table A2.39: Active mines (based on the US Geological Survey).

Mineral	Major operating companies and major equity owners	Location of main facilities	Annual capacity (10^3 tons)
Aggregate	ARC Ltd. (Hanson Plc., 100%)	50 quarries in various locations	50,000
	Foster Yoeman Ltd.	Glensanda quarry at Oban	15,000
Ball clay	Watts, Blake, Bearne & Co. Plc.	Various operations in northern and southern Devon	500
China clay (kaolin)	ECC Group Plc.	Mines and plants in Devonshire and Dorsetshire	3,000
Coal	RJB Mining Plc.	19 mines in various locations	40,000
Fluorspar	Durham Industrial Minerals Ltd.	Mines in Weardale	50
Fluorspar	Laporte Industries Plc.	Mill at Stoney Middleton, Mines in Derbyshire	70
Gypsum	British Gypsum Ltd.	Mines in Cumbria, Nottinghamshire, and Sussex	3,500

continued

Table A2.39: *Continued.*

Mineral	Major operating companies and major equity owners	Location of main facilities	Annual capacity (10^3 tons)
Potash	Cleveland Potash Ltd.	Boulby Mine, Yorkshire	500
Salt, rock	Imperial Chemical Industries Plc.	Mines at Winsford, Cheshire	3,000
Salt, rock	Irish Salt Mining and Exploration Co.	Carrick Fergus, Northern Ireland	300
Sand and gravel	TMC Pioneer Aggregates Ltd.	Chelmsford, Essex	1,000,000
Silica, sand	Hepworth Minerals and Chemicals Ltd.	Operations in Cambridgeshire, Cheshire Humberside, and Norfolk	6,000
Slate, natural	Alfred McAlpine Slate Ltd.	Penrhyn quarry, Bethesda, North Wales	25
Talc	Alex Sandison and Son Ltd.	Unst, Shetland Islands	15
Talc	Shetland Talc Ltd. (Anglo European Minerals Ltd., 50%; Dalriada Mineral Ventures Ltd. 50%)	Cunningsburg, Shetland Islands	35
Tin, ore	Crew Group of Canada	South Crofty Mine, Cornwall (Closed March 1998)	1,800

A2.15.2 Inactive mines

Some inactive underground mines in the UK are presented in Table A2.40.

Table A2.40: Inactive underground mines in the UK.

Name	Location	Mineral exploited	Dates of operation
Annesley-Bentinck	Near Kirkby, Nottinghamshire	Coal	Closed in 2000
Silverdale (deep mine)	Staffordshire	Coal	Closed in 1998
South Crofty	Redruth, Cornwall	Tin	Closed in 1998
Frazers Hush	Rookhope/North Pennine	Fluorspar	Closed in 1998–1999
Groverake	Rookhope/North Pennine	Fluorspar	Closed in 1998–1999

www.ingramcontent.com/pod-product-compliance
Ingram Content Group UK Ltd.
Pitfield, Milton Keynes, MK11 3LW, UK
UKHW020249290926
15470UKWH00002B/829